Ecology and Control of Introduced Plants

The global spread of plant species by humans is both a fascinating large-scale experiment and, in many cases, a major perturbation to native plant communities. Many of the most destructive weeds today have been intentionally introduced to new environments where they have had unexpected and detrimental impacts. This book considers the problem of invasive introduced plants from historical, ecological, and sociological perspectives. We consider such questions as 'What makes a community invasible?' 'What makes a plant an invader?' and 'Can we restore plant communities after invasion?' Written with advanced students and land managers in mind, this book contains practical explanations, case studies and an introduction to basic techniques for evaluating the impacts of invasive plants. An underlying theme is that experimental and quantitative evaluation of potential problems is necessary, and solutions must consider the evolutionary and ecological constraints acting on species interactions in newly invaded communities.

JUDITH H. MYERS is a Professor in the Faculty of Agricultural Sciences and the Department of Zoology at the University of British Columbia, Canada.

DAWN R. BAZELY is an Associate Professor in the Department of Biology at York University, Ontario, Canada.

ECOLOGY, BIODIVERSITY, AND CONSERVATION

Series editors
Michael Usher   *University of Stirling, and formerly Scottish Natural Heritage*
Denis Saunders   *Formerly CSIRO Division of Sustainable Ecosystems, Canberra*
Robert Peet   *University of North Carolina, Chapel Hill*
Andrew Dobson   *Princeton University*

The world's biological diversity faces unprecedented threats. The urgent challenge facing the concerned biologist is to understand ecological processes well enough to maintain their functioning in the face of the pressures resulting from human population growth. Those concerned with the conservation of biodiversity and with restoration also need to be acquainted with the political, social, historical, economic and legal frameworks within which ecological and conservation practice must be developed. This series will present balanced, comprehensive, up-to-date and critical reviews of selected topics within the sciences of ecology and conservation biology, both botanical and zoological, and both 'pure' and 'applied'. It is aimed at advanced final-year undergraduates, graduate students, researchers, and university teachers, as well as ecologists and conservationists in industry, government and the voluntary sectors. The series encompasses a wide range of approaches and scales (spatial, temporal, and taxonomic), including quantitative, theoretical, population, community, ecosystem, landscape, historical, experimental, behavioural, and evolutionary studies. The emphasis is on science related to the real world of plants and animals, rather than on purely theoretical abstractions and mathematical models. Books in this series will, wherever possible, consider issues from a broad perspective. Some books will challenge existing paradigms and present new ecological concepts, empirical or theoretical models, and testable hypotheses. Other books will explore new approaches and present syntheses on topics of ecological importance.

Ecology and Control of Introduced Plants

JUDITH H. MYERS
DAWN BAZELY

CAMBRIDGE
UNIVERSITY PRESS

PUBLISHED BY THE PRESS SYNDICATE OF THE UNIVERSITY OF CAMBRIDGE
The Pitt Building, Trumpington Street, Cambridge, United Kingdom

CAMBRIDGE UNIVERSITY PRESS
The Edinburgh Building, Cambridge CB2 2RU, UK
40 West 20th Street, New York, NY 10011-4211, USA
477 Williamstown Road, Port Melbourne, VIC 3207, Australia
Ruiz de Alarcón 13, 28014 Madrid, Spain
Dock House, The Waterfront, Cape Town 8001, South Africa

http://www.cambridge.org

First published 2003

Printed in the United Kingdom at the University Press, Cambridge

Typeface Bembo 11/13 pt   *System* LATEX 2_ε [TB]

A catalog record for this book is available from the British Library

Library of Congress Cataloging in Publication data

Myers, Judith H., 1941–
Ecology and control of introduced plants / Judith H. Myers, Dawn Bazely.
    p.   cm. – (Ecology, biodiversity, and conservation)
Includes bibliographical references (p.).
ISBN 0 521 35516 8 (hb : alk. paper) – ISBN 0 521 35778 0 (pb : alk. paper)
1. Invasive plants.   2. Invasive plants – Ecology.   3. Plant invasions.   4. Plant conservation.
I. Bazely, Dawn, 1960–   II. Title.   III. Series.
SB613.5 .M94   2003   639.9'9 – dc21   2002033351

ISBN 0 521 35516 8 hardback
ISBN 0 521 35778 0 paperback

We dedicate this book to our families
Jamie, Isla and Iain
Peter, Madeleine, and Carolyn

Contents

Preface

The concept for this book goes back at least 15 years. In the meantime invasive plant species have become the 'flavor of the month' and the literature bursts with interesting new papers. Writing this book has been an exciting undertaking. We have written for a wide audience, and therefore take the chance that it falls between the interests of a variety of readers. Some may find sections to be too anecdotal. Others may find parts to be too technical. As we wrote we could not resist including some of the fascinating stories of the involvement of individuals in spreading plants. It is a scary thought that others may be introducing weeds of the future as we write. We hope that land managers who are charged with controlling invasive weeds and restoring habitats will find this book useful. We admire your efforts in tackling such complex problems. We value the great scientific and management contributions made by our colleagues in biological control, and are sorry we could not include all of the ideas and successes. For students, the experts of the future, we hope that invasion ecology and biological control stimulate your interest. There are many hypotheses to be tested and problems to be solved at this interface between basic and applied ecology. To all, we would be happy to get your feedback.

Many people helped in this project and two in particular deserve enormous thanks. First, Jamie Smith used up several red pens worth of ink editing the manuscript. His insights and suggestions have been invaluable. Second we thank Dawn's mother who provided weeks of babysitting. We also thank Madeleine, Carolyn and Peter Ewins, for letting Dawn come west to work on the book. Jenny Cory, Anne Miller, and Diane Srivastava read sections of the manuscript and made many useful suggestions. Isla Myers-Smith read and commented on the total manuscript and also prepared several of the figures which was extremely helpful. Charley Krebs has been a valued mentor. He facilitated this project by writing useful books on ecology and ecological techniques and by providing the laptop on which the book was written. We also thank our graduate students who have suffered too long from inattentive advisors. Jill Sutcliffe and Keith Kirby of English Nature, and Norman Yan and Sohail Zaheer of York University provided useful information and Dawn's plant ecology class provided valuable feedback which we appreciated. Mansour Mesdaghi from University of Agricultural Sciences and Natural Resources, Iran, kindly shared his laboratory manual on vegetation measurement.

Background maps used in some of the figures were obtained from the following sites: http://www.scottforesman.com/educators/maps/worldmap. html World Map http://baby.indstate.edu/gga/gga_cart/basewi.gif Wisconsin Map and Viso 4.0 for windows

We dedicate this book to our families
Jamie, Isla and Iain
Peter, Madeleine, and Carolyn

1 · *Introduction*

This book is about plants that have been introduced to new areas, usually new continents. It is about the attempts that have been made to characterize which introduced species will become serious weeds, about their impacts on native communities, and on how introduced weeds might be controlled. Studies of invasive plants have provided a rich literature in applied plant ecology. The invasion of plants into a new environment is an example of succession in action and an experiment on the role of species in communities. In the following chapters we bring together theory and application to focus on both what the study of introduced plants reveals about ecological processes and what ecological processes might be applied to management programs. We consider how community and population ecology can be brought to bear on the topic of invasive plants. Because many introduced plants become invasive and are considered to be weeds, we start here by defining weeds. Next, in this introductory chapter, we describe the socio-economic context surrounding introduced plants and introduce topics to be discussed in more detail in following chapters.

Weeds and the value of native species

Weeds are plants growing out of place or plants whose value has not yet been discovered. They have demanded the time, attention and resources of farmers, gardeners and proud homeowners for centuries. Despite being pulled, sprayed, cursed and competed with, in the long run the weeds always seem to come back. Weeds are plants that can grow at high population densities and can have a negative impact on other plants valued by humans. Many of the most dominant weeds in the world are those that have been introduced to new habitats, sometimes accidentally, but all too often intentionally, for what seemed to be a good idea at the time. The story of gorse in New Zealand is an example (Box 1.1). Colton and Alpert (1998) surveyed the public and found that while most were aware of what weeds are, few understood the ecological and economic

Box 1.1 · *Gorse in New Zealand, a country filled with invaders.*

Gorse, *Ulex europeaus*, currently a curse in New Zealand, was intentionally brought from England to be used for hedgerows. A farmer by the name of McLean was put in charge of spreading the gorse seeds to other landowners because he was considered to be responsible, and could be trusted to spread the seeds fairly among the farmers. Gorse hedgerows in New Zealand probably did keep sheep in their pastures, and might also have boosted the morale of homesick immigrants who remembered gorse from the British countryside. In hindsight, if McLean had realized that gorse would eventually escape from the hedgerows and come to dominate in many areas of the countryside (Hill *et al.* 2000), he as a responsible farmer would have refused to take part in the seed distribution.

Like many other plants that were transplanted to new countries without their natural enemies, gorse was able to outcompete the native vegetation. Introduced domestic herbivores such as cattle, sheep, and goats, are likely to have aided in this process. Gorse and grazing mammals had been interacting for generations in the Old World, and the spines on gorse make it unacceptable to most grazers. On the other hand, before sheep arrived in New Zealand, the vegetation had only to deal with browsing by moas (Aves: Dinorthithoformes), large, flightless birds, now extinct (Myers and Bazely 1991), and one or two smaller species of herbivorous birds. Therefore the introduced mammals are likely to have preferred the less well-defended, native plant species.

Since the European settlement of New Zealand, over 80 species of vertebrates have been introduced and have caused the extinctions of native plants and animals. However, the number of introduced mammals pales by comparison to the 1600+ species of plants that have been introduced. These have caused no known extinctions, but they have altered native ecosystems by outcompeting or smothering native species and altering successional pathways (Clout and Lowe 2000). The New Zealand Government today considers introduced species to be as much of a threat to national security as those that are normally considered in this category, such as invading armies of humans! While New Zealand has perhaps received more than its share of new plant species, this situation is not unique.

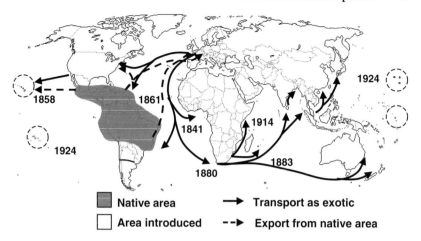

Native area   → Transport as exotic

Area introduced   --► Export from native area

Figure 1.1. Map showing years of transport and introductions of the shrub *Lantana camara*, around the world, from its native continent of South America. Modified from Cronk and Fuller (1995).

impacts of biological invasions by plants. This is not surprising because these impacts are varied and often complicated.

Not all weeds were introduced from faraway places. Indeed, many weeds are members of the native flora in an area. Similarly, not all introduced plant species that become established have obvious negative impacts on plant communities. For example, Arroyo *et al.* (2000) estimated that of 690 non-native plant species in Chile, 430 are considered weedy, compared with 132 native weeds. While there has been increasing concern over the consequences of human-aided transport of species around the world over the last 20 years, botanists have been well aware of the presence of plant species introduced from other countries for over a century (di Castri 1989). As early as 1905 Stephen Dunn published the *Alien Flora of Britain*, which examined introduced species in that country. In Australia, Robert Brown noted 29 species of introduced plants from Europe during a botanical reconnaissance of the coast near Sydney from 1802 to 1804 (Groves 1986b). Australia received its first European settlers in 1788. A good example of a plant that has been spread around the world is *Lantana camara* (Figure 1.1). This plant was valued as an ornamental. Its ability to invade and dominate new environments was not recognized until after it had become widely spread.

The history of plant introductions has been recorded by the activities of dedicated botanists who collected, pressed, mounted and archived plants in herbaria. This work has turned out to be a goldmine for ecologists

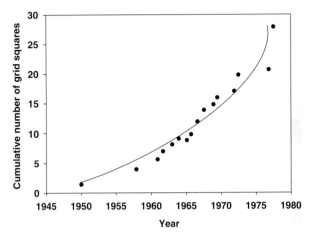

Figure 1.2. The spread of *Carduus nutans* in Australia as quantified by Medd (1987) from herbarium specimens. The spread was recorded in 0.5° longitude × 0.5° latitude squares. After Cousens and Mortimer (1995).

seeking to understand the patterns of spread of introduced plant species (Mack 2000) (Figure 1.2). There can, however, be inaccuracies associated with these records. While really different looking plant species are obvious and rapidly identified, the arrival of new species that resemble native species may be slow to be recognized. This was apparently the case for *Anthemis austriaca* in Pullman, Washington, USA, which was originally identified as the well-known species, *Anthemis cotula* (Mack 2000). Often the exotic nature of a species is indicated by its high density. Nevertheless, herbarium records can give a picture of the history of the expansion of the ranges of plant species following their introductions to other countries and regions.

In Europe historical records provide such detailed knowledge of plant movement that introduced species can generally be classified according to whether they arrived before or after the year 1500, although this date should more correctly be 1492, the year of return of Christopher Columbus from the New World (Scherer-Lorenzen *et al.* 2000, Williamson 2002). Given this long history of the spread, movement and naturalization of plant species in foreign locations (di Castri 1989), we might question why there has been so much recent interest in this topic. There is even a question of how native species should be identified. For example, if plants occurred in an area before the ice age, and do not currently exist there but have recently been reintroduced, are

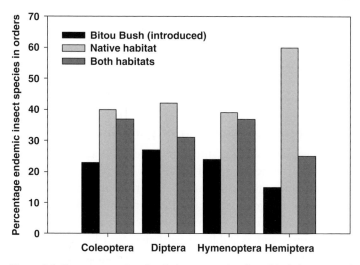

Figure 1.3. Percentage of endemic insect species found in habitats invaded by Bitou Bush, *Chrysanthemoides monilifera*, and native coastal habitats in New South Wales, Australia. Specialist insect species may be more affected in the areas invaded by the exotic species. Data from French and Eardley (1997).

they native or exotic? In defining native species the particular environmental and climatic conditions that currently influence their distribution can be important.

Native species are usually valued above introduced species by ecologists and environmentalists. However, for horticulturalists, introduced ornamentals are often given equal or higher preference. The value of exotic versus native species is discussed by Kendle and Rose (2000) who list five common arguments that are frequently used to support the protection of native over introduced species in landscape plantings. First natives may be more hardy than exotics and better adapted to the local environment. On the other hand, exotics may become invasive and outcompete natives. Additional arguments for the advantages of native plant species are that (1) they support more associated species such as insect herbivores (Figure 1.3), (2) their genetic diversity represents unique adaptations that should be protected from contamination through gene flow from exotic species, and (3) native plants define the landscape character. However, Kendle and Rose (2000) conclude that much of the research behind these arguments is either limited or equivocal. They also conclude that many of the justifications for using native plant species in landscape plantings have more to do with societal values than science. It is interesting that many plant introductions arose from Europeans attempting to recreate a familiar landscape character after they colonized new continents (Box 1.1). One of the main objectives of this book is to consider from an ecological perspective whether native plants should be valued over introduced plants.

The socio-economic background of plant introductions

Human-aided movement of plant and animal species is an intrinsic part of our history and social development (di Castri 1989) (see Chapter 2). Many of the major crop and domestic species that sustain the human population have been introduced species (Pimentel *et al.* 2000) and have clear benefits to the human race. In addition to crops, many introduced plants are ornamentals and valuable to gardeners and horticulturalists. The horticultural industry in the United States of America is worth billions of dollars – in 1998 US consumers spent $8.5 billion on lawn and garden supplies (McCartney 1999). A Gallup poll conducted in February 1999 revealed that many gardeners are middle-aged, highly educated homeowners with 'impressive' household incomes (McCartney 1999). In the USA, gardening has ranked in the top five favorite leisure-time activities in the six Harris polls conducted from 1995 to 2001, and in 1999, it ranked third (15%) behind reading (27%) and watching TV (22%) (Taylor 2001). C. D. Andrews, executive director of the Canadian Nursery Landscaping Association remarked in 2001, that gardening is the fastest growing leisure activity in North America (Porter 2001). This has created a public demand for new and exciting species (Mack and Lonsdale 2001) and international trade in plants and plant seeds.

One example of the international market in plants that we have observed is the sale of seeds of many Eastern North American prairie species at a local garden festival in Wales, UK. Another example was highlighted in a newspaper article in Vancouver extolling the creativity of a nursery owner who had imported a plant species from China to Canada that 'appears to be unaffected by pests or disease and does not resent being sheared'. Surely these are the characteristics of a species that will be hard to stop once it moves from the gardens into parks. Continuing in this 'shouldn't we be concerned' vein is an analysis of 'native' wild flower seeds in the Pacific Northwest carried out by Lorraine Brooks and Susan Reichard, University of Washington, Seattle. They found that seed packets from nine companies contained from three to 13 invasive plant species with seeds from one company containing nothing but introduced and potentially invasive plant species. It seems that seed companies are confused about the definition of native species.

In addition to the issue of introduced plant species the growing popularity of gardening has other negative impacts on the environment. In the United Kingdom, widespread mechanical extraction of peat from moorland bogs, for use by gardeners, threatens a number of plant communities that have been designated as 'Sites of Special Scientific Interest' by English Nature. The negative impact of gardening's popularity was even mentioned by the Member of Parliament, Caroline Flint, in a debate in the British House of Commons on 3 November 1999 (Hansard 1999). She stated that a number of gardening companies 'have turned large parts of Hatfield moors into nothing more than a lunar landscape'. Ms Flint pointed to the need for British celebrity TV and radio gardeners to take a stand on the issue. The growing popularity of gardening in Britain has led in recent years to a fourfold increase in the use of peat. With the movement of peat comes the distribution of plant seeds to new regions.

Billions of dollars are spent controlling weeds, pests and pathogens world-wide, many of which are introduced species. Pimentel *et al.* (2000) have estimated the annual cost of controlling and managing some non-indigenous species to be $136 630 million. Costs of control can easily be totaled in monetary values, but there are other less easily quantifiable costs. Some introduced species are less of a problem in agricultural areas, and more of a problem for natural communities. Here they may affect the provision of 'ecosystem services', by modifying water tables, reducing water flow in streams, and modifying wildlife habitat (Zavaleta 2000a). Naylor (2000) discusses the complexity of determining the costs of introduced species, since cost–benefit analyses are based on human values. It is difficult to estimate the values to place on non-use and indirect use of ecosystems (Chapter 3).

There are increasing calls for us to change our habits in terms of how we treat our environment. One example of this is the US Environmental Protection Agency's promotion of 'Beneficial Landscaping' (EPA 2002). The rationale for practicing beneficial landscaping includes, among other reasons, the possibility of reversing biodiversity loss due to 'clearing of native habitats and the introduction of invasive exotic plants'. The general principle of beneficial landscaping is that it is possible for people to meet their 'needs and sense of beauty, while maintaining or restoring healthy natural ecosystems'. These principles are outlined in Table 1.1. As discussed previously, Kendle and Rose (2000) question the rationale

Table 1.1. *Principles of beneficial landscaping that can be applied in any area experiencing human activity*

Principles of beneficial landscaping
Protect existing natural areas
Select regionally native plants
Reduce use of turf
Reduce or eliminate use of pesticides
Compost and mulch on-site
Practice soil and water conservation
Reduce use of power landscape equipment
Use planting to reduce heating/cooling needs
Avoid use of invasive exotics
Create additional wildlife habitats

From EPA (2002).

behind selecting natives and avoiding exotic plants in landscape design. Clearly the EPA principles go beyond a focus on introduced plants to incorporate a broader range of ecologically friendly changes in landscaping procedures. However, the conflict remains between the ecological view that exotic species are bad, and the view of horticulturalists that exotic plants are, if not good, at least not demonstrated to be bad.

Turning back the clock – is restoration possible?

Most introduced species are not invasive but those that are can have a spectacular impact when they come to dominate large areas. For a number of these species the initial stages of population increase and spread will appear to be slow (Figure 1.4). This early phase represents what Naylor (2000) calls a 'time-bomb'. She also discusses the long-term benefits of early control as compared to dealing with the problem following widespread establishment. A very serious problem is that humans only seem to respond strongly to perceived crises. This leads to a need to create an atmosphere of urgency in regard to preventing the introduction of new species and to attacking those species with a clear potential for invasion (Myers *et al.* 2000).

Applied biologists and land managers are challenged by the need to conserve the diversity and integrity of natural plant communities in the face of invasive introduced species, increased disturbance and changing

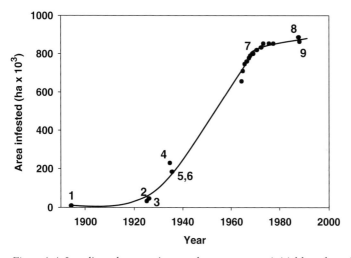

Figure 1.4. Invading plant species can demonstrate an initial lag phase followed by a log phase of population growth as demonstrated here by the cactus *Opuntia aurantiaca* in South Africa. This plant was introduced in the 1840s and established in glasshouse collections. The rapid spread between the 1930s and 1964 was due to both real population spread and increased search efforts. During this period of spread control efforts were applied including mechanical control, herbicide treatment and biological control by the cochineal insect, *Dactylopius austrinus*. The latter has reduced the density of the cactus but not its distribution. Numbers refer to different studies carried out between 1892 and 1981. After Moran and Zimmerman (1991).

environmental conditions. To meet this challenge effectively, three basic ecological interactions must be recognized: plant–environment, plant–plant, and plant–herbivore. The ability to restore a habitat to its original state is a litmus test of our ability to put ecological principles into action. Long-term quantitative assessment and well-designed experiments allow the evaluation of what works and what does not. In this book we appeal for more quantitative studies of invasive species and their control. Thus we include in an Appendix an introduction to population and biodiversity measurement techniques.

The introduction of non-native plant species to ecosystems and habitats can be considered as grand experiments. The addition of new species allows us to study the roles of species in these communities (see Chapters 3, 4 and 7 in particular). It provides us with an opportunity to assess whether ecological theory is able to predict pathways of change in species abundance and plant community composition. For the manager challenged with habitat conservation, the most important question is whether the

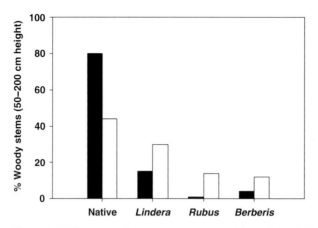

Figure 1.5. Changes in the percentage composition of woody browse species following reductions of deer populations from >54 deer km^2 prior to 1993, to 7–9 deer km^2 from 1993 to 2001 in a southern Ontario deciduous forest. Black bars are species' occurrences before deer removal and white bars are from 2001. Sample sizes were $n = 247$ stems in 1994 and $n = 403$ stems in 2001. Plant groups are palatable native trees and shrubs, and less palatable species – spicebush, *Lindera benzoin*, bramble, *Rubus idaeus*, and introduced Japanese barberry, *Berberis thunbergii*. (S. Chopra and D. R. Bazely, unpublished data.)

impacts of introduced species can ever be reversed or if, once an exotic is established, the community reaches a new 'equilibrium state'. Disturbed habitats are frequently assaulted by several factors simultaneously. Increased browsing may facilitate the invasion of a well-defended plant. However, the establishment or dominance of such a species may not be reversed by the removal or reduction of the browser population (Figure 1.5).

The irreversibility of a modified plant community after removal of a disturbance is shown in Rondeau Provincial Park, a deciduous forest in southern Ontario, Canada. In this case high densities of deer selected for increases in plant species defended against browsing and many native understory species disappeared from the forest plant community (Koh *et al.* 1996). The managers challenged with restoring the original plant community reduced the numbers of white-tailed deer, *Odocoileus virginianus*, in 1993 from over 400 to fewer than 100. However, following reduced deer densities, the unpalatable woody species continued to increase significantly. These were spicebush, *Lindera benzoin*, a highly aromatic, small, native tree; the prickly, native bramble, *Rubus idaeus*; and the thorny, non-native species, Japanese barberry, *Berberis thunbergii*, a cottage garden escape. This result ran counter to predictions that the regeneration of native, palatable tree and shrub species would increase with reduced deer browsing. The assumption is that the defended plants benefited more from deer removal than undefended, native species. Reduced seed banks of native species could also have been associated with poor regeneration (see Chapters 4 and 5). This is not the result a manager would have hoped for.

For the land manager, restoration is complex and unpredictable (see Chapter 9). While eradication of introduced species may be considered, the reality is that once a species has reached the point where it is recognized as a problem, eradication is not likely to be achievable even with massive economic investment. There have been successful eradications of introduced animal species and examples include rats, *Rattus* spp., from islands (Myers *et al.* 2000), and nutria, *Myocaster coypus*, in the UK (Gosling and Baker 1989). Mack *et al.* (2000) cite two examples of successful plant eradications, but these two species were present in very localized areas. As research into the economics of introduced species develops, we may see more costly eradication programs. However, benefit–cost analyses of the type proposed by Naylor (2000) are rare. Thus, the more common option has tended to be to do nothing or to attempt to control the species to some acceptable level through chemical or biological methods (see Chapters 7 and 9).

Biological control as an approach to introduced weeds

If introduced weeds have a competitive advantage because they lack natural enemies, then re-establishing the natural enemy complex should be a way of leveling the playing field. It seems the obvious solution. Classical biological control is the introduction of natural enemies in an attempt to control exotic pests (Chapter 7). It is based on the premise that one, some, or all of the natural enemy complex can reduce the density of the introduced pest to the extent that native vegetation will be able to assert a competitive edge. In native habitats, even weedy species tend to have lower and often more patchy distributions than the same species in an exotic habitat. The hope is that introducing part or all of the natural enemy complex will cause the reduction of the plant density to something more similar to the native situation.

Biological weed control has been highly successful in some cases and the most spectacular of these cases are cited time after time in ecology textbooks. Two species of weeds that have been greatly suppressed through biological control are prickly pear cactus, *Opuntia* spp., and St John's wort, *Hypericum perforatum*, both projects undertaken in the first half of the twentieth century. More recent successes include tansy ragwort, *Senecio jacobaea*, in western North America and an aquatic fern, *Salvinia molesta*, in many tropical lakes (Room 1990). Although these and other clearly successful programs represent only a fraction of those that have been attempted, biological control is the only potential solution for many invasive weeds. It is an area of increasing effort. By the early 1980s, approximately 100 species of plants had been targeted for biological control and by early 1990s this number had doubled (Julien 1992, Julien and Griffiths 1998). High proportions of these weeds are in the families Asteraceae and Cactaceae.

Just as the introduction of plants to new communities can be viewed as ecological experiments, so can the introduction of biological control agents. These programs, if appropriately monitored can be used to test hypotheses on interactions among plants and their natural enemies. In Chapter 7 we consider the problem of predicting what characteristics make a good biological control agent and why some agents reduce the population density of their food plants while others do not. Biological control is often the best hope for managers faced with reducing the impacts of introduced species, and any increase in the ability to predict what makes a successful program will be greatly valued.

Promoting ecosystem management for native species

Thus far we have considered the negative attributes of weeds. However, native weeds are part of the agricultural environment that has changed dramatically in recent times. A group of weeds that are now declining are known as 'agrestals'. These are annual weeds found in arable fields and their seeds have become adapted to disperse with crop seeds (Kornas 1988). Research during the 1950s in the Western Carpathian Mountains of Poland showed that this was the major group of weeds in cereal and flax fields. By 1985, a number of these species of weeds had disappeared from the fields (Kornas 1988). This has happened in part for two reasons – improved technology for sifting weed from crop seeds, and more sophisticated herbicides (Fox 1990). Indeed, a number of the 'old aliens' – introduced pre-1500 from southern and southeastern to central and northern Europe – are the subjects of conservation programs (e.g. *Adonis aestivalis*, *A. flammea*, and *Agrostemma githago*) (Scherer-Lorenzen *et al.* 2000). This story is an excellent example of how ecosystem-level practices, in this case in agricultural habitats, can alter the conditions so that particular species disappear.

The importance of ecosystem-level attributes such as disturbance regimes in determining the abundance and distribution of plants is increasingly being recognized both for rare and endangered species but also for non-native species. Understanding how alterations in the landscape, land use and ecosystems affect non-native species will help the manager mitigate habitat changes that may increase the likelihood of invasion (Chapter 9).

Conclusions

In this introduction, we raise some of the topics that we consider in the chapters to follow. It is clear that human activity has led to both the accidental and intentional global redistribution of plants. Therefore social, economic, and ecological influences are intimately involved in studies of introduced species. Humans will continue to move and more plant species will go with them and create a continuing need for the management of plant communities. However, the movement of plant species is also a valuable perturbation for the study of the ecological principles that are relevant to future management schemes as well as the understanding of plant communities. We will next look in more detail at the patterns, impacts and control of introduced plant species.

2 · *Planet of Weeds: exotic plants in the landscape*

In this chapter, we consider the magnitude and scope of the invasive species problem, the language that is used when discussing introduced species, historical patterns of introductions, and the dynamics of colonization and spread. We also examine interactions between introduced species, habitat fragmentation and disturbance, and briefly describe some ecological theory relating to invasive species. We begin with a consideration of the role of invasive plants in the global environmental changes that are currently occurring.

The biota of the earth is dynamic both in the short-term and long-term. As students of evolutionary biology are quick to learn, most species that have ever existed on the earth have become extinct. For example, it is estimated that 96% of all species extant or living 245 million years ago may have become extinct (Raup 1979). Speciation and extinction are fundamental outcomes of evolutionary processes. However, rates of extinction vary over time and there have been periods of 'mass extinction', which can be seen in the fossil record as the disappearance of large numbers of species (Benton 1995, Eldredge 1997).

In the last 25 years, beginning with Norman Myers (1976), many biologists and ecologists have adopted and echoed the refrain that we are presently in such a period of accelerated extinction. But, unlike past episodes of mass extinction, this one is directly caused by human activities and has been termed the 'Biodiversity Crisis' (see definitions in Box 2.1) (Wilson 1985, Wilson and Peters 1988). Estimates of current extinction rates are estimated to be 10 to 100 times greater than those of past extinction episodes (Lawton and May 1995). The five main causes of human-induced extinctions, originally defined by Diamond (1989), are: habitat destruction, habitat fragmentation, over-exploitation, **introduced species** (see definitions in Box 2.2) and secondary effects or 'chains of extinction' cascading through an ecosystem. To this list can be added pollution (Lande 1999) and global climate change (Chapin *et al.* 2000). All of these human activities or human-induced effects may be

---

Box 2.1 · *What is biodiversity?*

Biodiversity is a ubiquitous term. We see it everywhere from newspaper headlines to television reports to ecology textbooks (but not in our computer spell checks). What exactly is meant by this term? In 1992 the Convention on Biodiversity was signed at the Rio de Janeiro Earth Summit, organized by the United Nations Environment Program. Article 2 of the Convention defines Biodiversity as: 'the variability among living organisms from all sources including, *inter-alia*, terrestrial, marine and other aquatic ecosystems and the ecological complexities of which they are a part. This includes diversity within species, between species and of ecosystems' (UNEP 1992).

Box 2.2 · *What is an introduced species?*

Species present in community assemblages are generally categorized as being either **native** – indigenous, or endemic, including those resulting from prehistorical invasions, or **introduced** (**exotic**) – resulting from historical invasions including natural range extensions and human-mediated introductions. In Europe, species are sometimes categorized according to whether they arrived *before* (**archaeophytes**) or *after* (**neophyte** or **kenophyte**) the year 1500 (Kornas 1990). Carlton (1996) has pointed out that a number of species may not clearly fit into any of these categories. He has defined species that are not clearly known to be either native or introduced as **cryptogenic**. Without knowing the proportion of species that are cryptogenic it is difficult to evaluate invasion corridors, invasion success rates, or the susceptibility of communities to invaders. Defining species as being native or introduced will not always be simple, and cryptogenic species may be common in some plant communities.

---

responsible for declining species richness, reduced genetic variation within species or the deterioration of different types of ecosystems.

These factors do not act independently. For example, habitat fragmentation, by creating more edge habitat, may increase the likelihood that a plant community will be colonized by non-native species (Brothers and Spingarn 1992). These new species may have negative impacts on already vulnerable or reduced populations of native species. Climate change has the potential to increase or modify the ranges of both native and non-native species. It is predicted that harmful, introduced plant species in the United States will increase their ranges as a result of global warming (OTA 1993, Zavaleta & Royval 2002). Therefore, introduced species, in conjunction with a plethora of other changes, are likely to contribute to the biodiversity crisis.

Much of the evidence that introduced species cause extinction does not come from studies of introduced plants, but from those of introduced animals, generally predators, and plant diseases (Campbell 2001). Many of these studies involve animals on islands and, in particular, species of birds that have gone extinct following introduction of predatory species, such as the brown tree snake, *Boiga irregularis*, on Guam (Savidge 1987, Pimm *et al.* 1995, Steadman 1995). Extinction is difficult to document, and obtaining accurate estimates of extinction rates is made more difficult by a paucity of data (Regan *et al.* 2001).

Since 1600, five hundred and eighty-four plant species have been certified as having become extinct (Smith *et al.* 1993). What was the role of introduced species in causing these extinctions? The many difficulties associated with documenting extinction rates make it hard to find a single metric with which to measure extinction (Regan *et al.* 2001). Therefore it is useful to use a surrogate measurement. Smith *et al.* (1993) suggested that changes of species from *vulnerable* to *endangered* status in the Red Data Books, for example, could be used as an indicator of potential extinction rates. In the case of introduced plant species, a surrogate for estimating the likelihood of extinction might be a decline in overall plant species richness in areas with introduced species.

Wilcove *et al.* (1998) point out the lack of quantitative studies evaluating the threat posed by non-native species to native species. However, using species classified by The Nature Conservancy as possibly extinct, critically imperiled or imperiled, the Network of Natural Heritage Programs and Conservation Data Centers was able to identify the causes of threat to 1880 of the 2490 imperiled species. In their analysis they found that of 1055 imperiled plant species in the United States, over 81% were threatened by habitat destruction and degradation, and 57% were threatened by alien species. It should be noted that a species could be threatened for more than one reason. Other threats of less significance categorized by Wilcove *et al.* (1998) included pollution, over-exploitation and disease. It is often unclear whether these threats were determined using qualitative or quantitative criteria, or simply by 'expert' opinion. However, this analysis gives some indication of the magnitude of the impacts from introduced species.

The scope of the problem: how many and how costly are non-native plant species?

There are no global estimates of the numbers of non-native plants, but national and regional estimates exist. Approximately 28% of plant species in Canada are currently estimated to be non-native or introduced (Heywood 1989), and nearly 50% of the vascular plants in New Zealand are non-native (Green 2000). Overall, estimates vary hugely from area to area depending in part on how well the flora and plant communities have been described and studied (Table 2.1). Not surprisingly, there have been calls for the establishment of global databases that coordinate the monitoring of introduced species (e.g. Ricciardi *et al.* 2000).

The estimated cost of all invasive species in the United States is $125 billion annually (Baker 2001), and introduced weed species account for approximately $23.4 billion in annual crop losses (Pimentel *et al.* 2000). Invasive plants occupy over 100 million acres (1 acre = *c*. 0.40 ha) in the United States and are estimated to be spreading at the rate of 3 million acres per year (National Invasive Species Council 2001). The enormity of the problem has resulted in various laws being passed in the United States over the last 15 years that call for monitoring the spread of invasive plant species. These have culminated in a Presidential Executive Order that instructs federal agencies to develop policies for invasive species on federal land and to form an Invasive Species Council (OTA 1993, Clinton 1999, Reichard and White 2001).

A major hurdle to developing legislation for invasive species, is that it could be viewed as conflicting with international agreements on free trade that restrict any regulations deemed to be impeding imports of commercial plants. Policies must be based on scientific principles and justified by risk assessments. The level of protection must be appropriate to the documented risks and cannot unduly restrict trade (Campbell 2001). In addition to the United States, many other countries, such as Canada, Australia, New Zealand, and the United Kingdom, have developed, or are developing, strategies and legislation for dealing with introduced species (Williams and West 2000, Young 2000, Maltby and Mack 2002). The most extreme example of a response to introduced species is that of New Zealand. There the government acted directly and vigorously, moving beyond calls for merely monitoring the spread of species.

Table 2.1. *Percentages of introduced plant species in countries around the world*

Continent/Country	Native	Introduced	% Introduced	Source
Oceania				
New Zealand	2400	2100	50	1
Australia	15 638	1952	11	17
Hawaii	956	861	47	3
Cook Islands	284	273	49	2
Fiji	1628	1000	38	2
French Polynesia	959	560	37	2
New Caledonia	3250	500	13	2
Solomon Islands	3172	200	6	2
North and Central America				
Canada	3200	881	28	15
British Columbia	2475	662	21	4
Alberta	1475	280	16	5
Ontario	2056	805	28	5
Quebec	1803	740	29	5
Devon Island	115	0	0	5
Prince Edward Is.	624	316	34	5
Newfoundland	906	292	24	5
United States and Caribbean				
Alaska	1229	144	11	5
California	4844	1025	18	5
Florida	2523	925	27	6, 16
Illinois	2058	782	28	5
Utah	2572	444	15	5
New Mexico	2680	229	8	5
Tennessee	2208	507	19	5
Texas	4498	492	10	5
Great Plains	2495	394	14	5
New England	1995	877	31	5
Virginia	2056	427	17	5
Bermuda	165	303	65	5
Bahamas	1104	246	18	5
Cuba	5790	376	6	5
Puerto Rico	2741	356	12	5
Belize	3023	107	3	5
South America				
Chile	4500	690	15	7

Table 2.1. *(cont.)*

Continent/Country	Native	Introduced	% Introduced	Source
Africa				
South Africa	*c*. 19 000	8750	46	18
North Africa		100–200	2–4	8
Europe & the Middle East				
Europe – Mediterranean		250	1	9
Spain		637	13	10
Israel		137	5	11
Germany	2850	480 (315)*	14 (9)*	12
British Isles	1350	558	29	13
France	4171	292	7	14
Norway	1396	148	10.6	14
Austria	2972	211	7.1	14
Switzerland	2789	304	10.9	14
Hungary	2462	325	13.2	14

*Total is both neophytes and archaeophytes, while neophytes (introduced since 1500) are in parentheses.

Sources: 1. Green (2000), 2. Vitousek *et al.* (1996) and citations therein, 3. Wagner *et al.* (1990), 4. Taylor and MacBryde (1977), 5. Rejmánek and Randell (1994), 6. OTA (1993), 7. Naylor (2000), 8. Le Floc'h *et al.* (1990), 9. Quezel *et al.* (1990), 10. Vila *et al.* (2001), 11. Dafni and Heller (1990), 12. Scherer-Lorenzen *et al.* (2000), 13. Williamson (2002), 14. Weber (1999), 15. Heywood (1989), 16. Gordon (1998), 17. www.anbg.gov.au/cpbr/australian-plants/statistics.html 18. http://fred.csir.co.za/plants/global/continen/africa/safrica/index.html

The New Zealand Biosecurity Act (New Zealand Government 1993) legislates the eradication or effective management of pests and unwanted organisms already in the country, as well as requiring that all non-native plants and other species be prevented from entering the country.

There is a tendency to treat all introduced organisms as being equally threatening. However, the impact of most non-native species is poorly known, although all species are clearly not equal in their impacts (Heywood 1989). Some plant and animal species move quickly to colonize and, once established, suppress or prey on native species. Other species spread very little, if at all, beyond their point of introduction. It is therefore incorrect to assume that all introduced species have potentially huge effects. Obviously, the response of New Zealand is a blanket approach that would be hard to enforce in countries that are not islands. A ban on plant introductions is difficult to enact in the face of the enormous

Table 2.2. *Names used in the literature to refer to foreign or native plant species*

Come from aways				Stay at homes
General	Temporary	Permanent	On the move	
non–native	casual	naturalized	colonizers	native
non–indigenous	persistent	established	invasives	indigenous
exotic			invaders	endemic
imported				
immigrant				
introduced				
aliens				
novel				
adventive				

After Davis and Thompson (2000), Clements and Foster (1994), Quezel *et al.* (1990).

commercial markets for ornamentals that are worth millions of dollars. Since biological invasions are usually irreversible, it is perhaps best to design programs around the slogans 'If in doubt, keep it out' and 'Guilty until proven innocent'. As will be discussed later, however, strict importation regulations on introducing exotic species can also restrict classical biological control programs that use introduced natural enemies to reduce population densities of invasive species.

What's in a name?

What should we call a species whose introduction to a new continent is mediated in some way by humans? A number of different terms have been used in the 'invasion ecology' literature and in floras, to indicate the foreign status of introduced species (Table 2.2). Some of these terms refer only to the fact that the species has come from afar, and is foreign. Other terms carry associated implications about the potential impacts of the species in the new habitat. Davis and Thompson (2000) describe eight ways to classify species that are moving into new areas. This classification is based on whether the species have moved long or short distances, are regionally common or novel, and whether their impacts on the local communities to which they have moved are large or small. In this system all species that are expanding their ranges into new areas are considered to be 'colonizers' but only those that have large impacts are called 'invaders'.

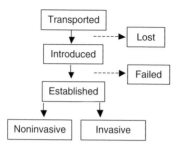

Figure 2.1. Stages in the introduction, establishment and spread of a non-native species.

In the United Kingdom, the terms used to categorize introduced plant species are associated with their ability to colonize and spread (Clements and Foster 1994). 'Naturalized' species are those that have become established to the point where they appear to be native species. 'Established' plant species are introduced species that are on their way to becoming naturalized, and are a permanent feature of a plant community. These species are distinct from 'casual' and 'persistent' species. **Casual** species are those that will not persist in a locality for more than two years without reintroduction, while **persistent** species may last longer than two years, but are not reproducing successfully, so do not have the possibility of becoming permanently established.

A problem with classifying introduced species based on their real or anticipated impact is actually measuring the 'impact' at the level of the individual plant (see later chapters). This concern led Daehler (2001) to suggest that the term 'invader' be restricted to novel species that spread in a new environment. If these species reach high densities as they spread, it is likely that they will have some ecological impacts in the native communities even if it involves only using space. Measuring the invasion of a species into new areas should be more straightforward than quantifying its impacts. Unfortunately, invasions of new species are not frequently monitored or quantified.

The most accurate use of the term 'invasive species' would be to describe a species introduced from a different area, most often a different continent, which first becomes established (colonizes), increases in density and expands rapidly across a new habitat (Figure 2.1). Many of the other terms (Table 2.2) refer to the foreign source of the species with no implications about their dynamics. Some of these terms have other

Box 2.3 · *What is a noxious weed?*

Many political jurisdictions (countries, states, provinces) have 'noxious weed lists'. The definition of noxious differs among jurisdictions but may mean (1) that these plants are prohibited from being imported, (2) that they have the potential to cause harm and therefore are of economic importance, or (3) that they must be controlled by the landowner. In some cases very common and environmentally detrimental weeds are not put on noxious weed lists because an effective and economically feasible control procedure is not available. Environmental or 'community' weeds (*sensu* Fox 1990), those whose main impact may not be directly on agricultural or grazing lands but on natural areas, are often not placed on these lists. In some cases, a particular species may be on some noxious weed lists but not on others. One example of this is the mismatch between the provinces in which purple loosestrife, *Lythrum salicaria*, occurs in Canada and those where it is on the noxious weed list (Figure 2.2).

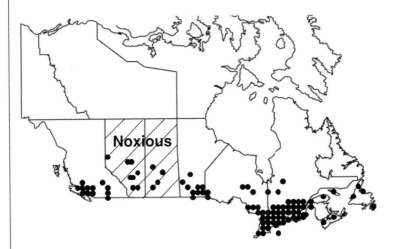

Figure 2.2. Dots indicate the distribution of purple loosestrife in Canada and hatching indicates provinces in which it has been declared a noxious weed. This shows that there is little correspondence between weed density and its declared status. Data from http://www.magi.com/~ehaber/factpurp.html

connotations that may cloud their interpretation. One of these terms is 'aliens', which has been tainted by its use in the science fiction literature to mean strange and often threatening. Human immigrants to new countries are also called aliens and, particularly in cases of illegal immigration, this term can have negative political overtones. Some feel that a negative reaction to introduced plants is exacerbated by using the term aliens and that this reaction is akin to racism (Kendle and Rose 2000). Similarly, 'exotic' is another term in common usage for introduced plants, and this also implies difference from the norm. It is important to remember that all species have the potential to expand their range – with or without human intervention – and that many nuisance species may have been **established** or even **naturalized** a long time before they become invasive. Many **naturalized** species are not seen as being overly abundant or damaging enough to cause them to be placed on lists of noxious weeds (see Box 2.3 on the classification of noxious weeds) or to become targets for control. In this book we have used 'introduced', 'non-indigenous' and 'exotic' to indicate plants that have been brought to a new area and 'invasive' for introduced plants that have become very common or are spreading. We feel that the terms alien and exotic can still be used to describe introduced plants.

Patterns of plant introductions

Plants have been moved around the world both accidentally and deliberately. It is sobering to realize that most weeds were intentionally introduced. Botanical gardens and nurseries have been excellent sources of introduced species (see Box 2.4 for an example of the role of nurseries on the naturalization of introduced plants) (Heywood 1989), and botanical gardens continue to trade seeds internationally. In the United States 82% of the 235 species of woody plant invaders were introduced as ornamentals or for landscaping (Reichard and Hamilton 1997). In the northeastern United States 60% of more than 600 naturalized taxa were introduced deliberately (Fernald 1950). Approximately 46% of the noxious weeds in Australia were introduced for ornamental or other purposes (Panetta 1993). In the city of Zurich, Switzerland, alone, 300 plant species have become established, of which 52% were planted for ornamental purposes. These invaders are thought to have led to the disappearance of more than 150 native species (Landolt 1993). A breakdown of the manner of introduction of naturalized plant species in South Australia is given in Table 2.3. It is clear that ornamentals top the list.

Table 2.3. *The manner of introduction of plants to Southern Australia*

Intentional	Documented	Suspected	Unintentional	Confirmed	Possible
Ornamentals	319	40	Attached to		
Fodder plants	58	17	stock	4	88
Culinary plants	43	1	Contaminated		
Hedges	14	0	seed	16	41
Medicinals	8	5	Ballast plants	7	36
Other	9	1	Footwear	0	11
			Contaminated		
			fodder	3	3
			Other	5	0
Total	451	64	Total	35	179

From Kloot (1991).

Box 2.4 · *The importation of nursery plants and their naturalization in Florida.*

Pemberton (2000a) compared 1884 non-native horticultural species listed as being sold by plant nurseries in Florida between the years 1886 and 1930 to the current vascular plant flora of Florida. He showed that 14% (264 species) have become naturalized. Plants that were sold for longer periods of time were more likely to have become established. Establishment was 2% for those sold for only one year, 31% for those sold from 10 to 30 years, and 69% for those sold for more than 30 years. Plants that had large ranges in their native habitats were more likely to establish than those with small native ranges, but there was no difference in the pattern of establishment between plants with close relatives in the native flora and those belonging to non-native genera and families. Vines and aquatic herbs had the highest rates of naturalization.

There is a fine tradition of introducing non-native plants to combat soil erosion. Examples of species introduced to the United States for this purpose include Russian olive (*Elaeagnus angustifolia*), multiflora rose (*Rosa multiflora*), kudzu (*Pueraria lobata*) and saltcedar (*Tamarix ramosissima*) (Box 2.5). It is not surprising that these plants, which were chosen because of their rapid growth and tenacious existence, would also become invasive.

In addition to terrestrial plants, many aquatic plants have been introduced through the aquarium trade. Invaders such as Eurasian milfoil, *Myriophyllum spicatum*, and *Salvinia molesta* have been introduced by people dumping aquarium water and plants down drains or into local bodies of water (Reichard and White 2001). The very aggressive aquatic weed *Hydrilla verticillata* was probably introduced this way (OTA 1993) and water hyacinth, *Eichhornia crassipes*, escaped from aquatic gardens in the southern United States to become a major problem by clogging waterways. In the United States more than $100 million are spent each year in controlling invasive aquatic weeds.

Just as humans introduce plant species to new areas, so they continue to be a major factor in their spread (Figure 2.3). Gardeners and horticulturalists frequently and intentionally spread ornamental plant species (Reichard and White 2001). Accidental spread of plants by humans is also a continual process as evidenced by the movement of aquatic weeds with boats. In New Zealand the spread and distribution of five submersed, vegetatively reproducing weeds among 107 lakes was associated with boating and fishing activities. Inspections showed that 5% of boats carried weed fragments of which 27% were from other lakes. The probability of spread decreased rapidly with distance and was very low beyond 125 km (Johnstone *et al.* 1985). The length of time that fragments of

---

Box 2.5 · *Introductions for land stabilization.*

A. Saltcedar (*Tamarix ramosissima*)
Tamarix ramosissima was initially introduced as a nursery plant in New Jersey in 1837. Although the records are poor, saltcedar was apparently planted along western streams to stabilize the banks. Between 1890 and 1920 it spread along many western rivers and by 1950 occupied most of the suitable habitat in the central part of the distribution, Arizona, New Mexico, and western Texas. From there it subsequently spread to surrounding states from Montana in the north to Mexico in the south. Six other species of the genus *Tamarix* are naturalized in the United States, but no others have become weedy. Saltcedar forms dense thickets and displaces native cottonwoods, *Populus fremontii*, and willows, *Salix gooddingii*. It also consumes vast quantities of ground water, compared with native vegetation, thus lowering water tables. Salt secreted onto tamarix leaves, drips down and increases soil salinity (De Loach andLewis 2000, Zavaleta 2000a).

B. Kudzu (*Pueraria lobata*)

Pueraria lobata came to the United States from Asia following its introduction at the Japanese Pavilion at the United States Centennial Exposition in 1876. Initially kudzu was valued for its rapid growth as a pasture plant and for provision of shade as a 'porch vine'. In the early 1930s soil conservation became a major focus of the United States government resulting in the Soil Conservation Act of 1935. This led to a variety of alien plants being introduced and widely planted for erosion control. Between 1935 and 1942 millions of kudzu crowns were planted in the southern United States. However, by the late 1940s farmers, foresters and highway engineers began to complain about kudzu engulfing trees and growing over buildings. Roadsides were blanketed with dense mats of kudzu several feet thick. In 1993 the Congressional Office of Technology Assessment estimated the economic costs of kudzu to be about $50 million annually. Currently the focus is on developing a biological control program for kudzu although some research has also been initiated to explore the potential of using kudzu fibers for paper and cloth.

The moral – the characteristics of plants that make them seem appropriate for erosion control are also those of invasive species (Kinbacher 2000).

http://www.cptr.ua.edu/kudzu/   http://www.sbs.uab.edu/history/Varticles/Kudzu4.htm

---

aquatic plants can survive out of the water will also affect their rate of spread among lakes. This varied among species in the New Zealand study from 45 hours for *Hydrilla* to 75 hours for *Ceratophyllum*. Tracking and predicting the patterns of spread of invasive species over large areas (see Boxes 2.6 and 2.7 for examples) is an applied problem in landscape ecology.

The European invasion

Europeans were particularly active in spreading plants as they colonized new parts of the world. Alfred Crosby (1986) recounts the history of the spread of plants to what he terms the 'Neo-Europes' – Canada, United States, Australia, New Zealand, Argentina, and South Africa. Regions in all of these countries have climates similar to Europe and therefore were suitable for both Europeans and their plants. Plants were brought

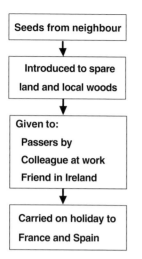

Figure 2.3. The spread of Himalayan balsam, *Impatiens balsamifera*, by Mrs Norris of Camberley, Surrey, UK (Rotherham 2000).

in for their medicinal, agricultural, and ornamental values. During the seventeenth century, plantain (*Plantago major*) was called 'Englishman's foot' by the Amerindians of New England because it 'grew where the English have trodden' (Crosby 1986). Thomas Jefferson, an early president of the United States of America, actively imported and spread Scotch broom from Britain to eastern North America, thus leaving this weed as a legacy for future generations.

In general, grasslands of the Neo-Europes had not evolved with grazing animals and, therefore, they rapidly shifted from being dominated by grazing intolerant native bunch grasses to grazing tolerant species, whose seeds spread with hay accompanying European cattle and sheep. As early as 1810, native grasses in Australia were in retreat, and by the mid-1800s, in the vicinity of Melbourne, Australia, 139 alien plant species were recorded. Of these species, 83 had also established in North America (Crosby 1986).

The origins of plants that are the targets of biological control programs give a picture of the non-random pattern of source areas for introduced plants that have become weeds (Figure 2.4). The European influence, measured as the proportion of introduced plants coming from that continent, has been considerably stronger in Chile than in areas that have a greater mix of temperate and tropical climates – Australia and South Africa. Australia seems to have given more weeds to South Africa than *vice versa*.

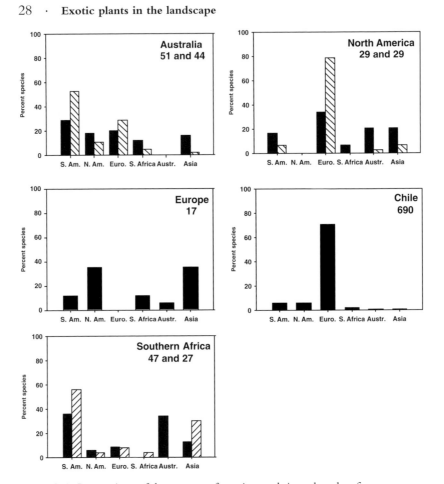

Figure 2.4. Comparison of the sources of exotic weeds introduced to four continents are based on lists of species in Cronk and Fuller (1995) (black bars) and weed species that are targets for biological control in Julien and Griffiths (1998) (hatched bars). Data for Chile are from Arroyo *et al.* (2000). Recipient areas and numbers of species are given.

Cronk and Fuller (1995) list a number of plant species that have been moved among continents. In some areas these plants have become invaders and in others they have only become established. This list is certainly not complete but it gives an indication of the trends of movement of plant species among continents (Figure 2.4). Patterns are rather similar to those based on weed control programs with North America receiving a majority of plants from Europe, Australia from South and Central

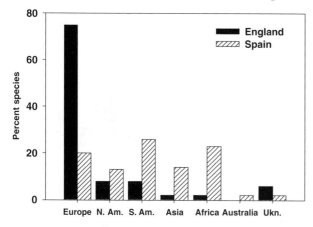

Figure 2.5. Origins of the introduced species of the flora of the City of Nottingham, UK (based on Rotherham 2000) and Spain (based on Vilà *et al.* 2001).

America as well as Europe and Eurasia, and South Africa receiving most plant species from South and Central America and Australia. Europe has received the most plants from North America and Asia. The pattern of plant introductions to Chile is very similar to that of North America (Arroyo *et al.* 2000).

Although Crosby (1986) emphasized the movement of plants to the Neo-Europes, a reverse flow has also occurred, although the number of species that have become seriously invasive may be less. In the city of Nottingham, UK, for example, many of the introduced species come from Europe (Figure 2.5) with fewer from North and South America. Another example of plant movement to England is illustrated by plant communities in the centre of the wool industry along the River Tweed, where 348 foreign plant species were recorded. Of these, 46% were European or Near Eastern in origin, 4% Asian, 14% Australasian, 6% North American and 12% South American. Most had seeds with spines or hooks (Harper 1965) and came with imported wool. The pattern of introduction of these 'wool aliens' to England can be compared with that of the European country, Spain (Figure 2.5). The sources of species introduced to Spain are similar to those of species introduced to Nottingham, but the contributions from source areas are distributed more evenly among other parts of Europe, the Americas, Asia, and Africa.

There are several examples of North American plants becoming invasive in Europe (Maillet and Lopez-Garcia 2000). One North American plant that is sufficiently detrimental to warrant a biological control program in Russia and Hungary is ragweed, *Ambrosia artemisifolia*. This species is a particular problem in agricultural areas (Reznik *et al.* 1994). The North American native goldenrod, *Solidago canadensis*, has been classified as an invasive in central Europe, including Germany (Cronk and Fuller 1995, Scherer-Lorenzen *et al.* 2000). *Rhododendron ponticum*, native to Turkey, Portugal and southern Spain, became invasive in the British Isles after it was introduced as an ornamental (Cronk and Fuller 1995, Burton 2000). Here it replaces the forest understory, and reduces the regeneration of native species. Sycamore trees, *Acer pseudoplatanus*, were also introduced from southern Europe to Britain where they are invasive. A plant from Asia that is a problem in both the United Kingdom and North America – Himalayan balsam, *Impatiens glandulifera*, has been spread far and wide by gardeners (Rotherham 2000). Thus, while Britain and Europe gave away many species, they have also received their share from other parts of the world.

Box 2.6 · *Illustrating spread – Eurasian water milfoil as an invasive species.*

One way to monitor the spread of a non-native species is to record the presence of the species in new locations defined by political boundaries such as counties or states. Eurasian water milfoil, *Myriophyllum spicatum* L. is native to Europe, Asia, and Northern Africa and was first identified through a voucher specimen in North America in Washington DC in 1942, although it may have been in the continent since the turn of the century (Buchan and Padilla 2000). Milfoil has now spread from the East to the West coasts of the United States (Figure 2.6). An example of the more local pattern of spread of milfoil is that recorded for lakes in different counties of Wisconsin (Figure 2.7). Although the availability of lakes in different counties undoubtedly varies, it is evident that this plant has spread from the southeastern corner of the state. The pattern of movement is not a simple diffusion process. Milfoil spreads by fragments being carried by boats and trailers as well as by birds and water. Therefore it is likely that introductions will have in some cases jumped to a new focus of spread by boaters visiting several lakes or birds migrating. The occurrence of milfoil in a particular lake in Wisconsin was determined more by the water quality there than characteristics associated with

Figure 2.6. Distribution of Eurasian milfoil, *Myriophyllum spicatum*, across the United States in 1950, 1965 and 1985. After Aiken *et al.* (1979), and Couch and Nelson (1985).

human activity (Buchan and Padilla 2000). This contrasts with the New Zealand study (Johnstone *et al.* 1985) described above (p. 26). At the lake level, the spread of an aquatic invader is more likely to follow a simple diffusion process, as in Lake George, New York (Figure 2.8).

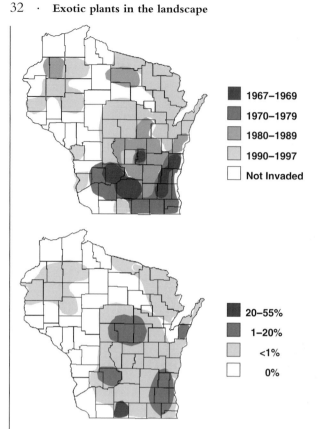

Figure 2.7. Invasion of Eurasian milfoil across the state of Wisconsin by year and as the percentage of lakes invaded in counties. After Buchan and Padilla (2000).

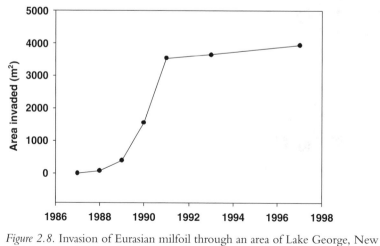

Figure 2.8. Invasion of Eurasian milfoil through an area of Lake George, New York between 1987 and 1997. Data from Charles W. Boylen.

Box 2.7 · *Illustrating spread – Japanese knotweed as an invasive species.*

Japanese knotweed, *Fallopia japonica*, was introduced to Britain in 1825 as an ornamental but has been described as the 'most pernicious weed in Britain' (Coleshaw 2001). Knotweed is a rhizomatous perennial plant and grows to heights of 2–3 m. Reproduction is primarily, if not exclusively, vegetative. Rhizomes can extend 7 m

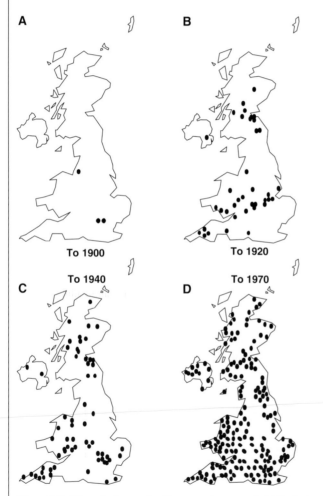

Figure 2.9. Historical spread of Japanese knotweed, *Fallopia japonica*, across the British Isles, (A) up to and including 1900; (B) up to and including 1920; (C) up to and including 1940; and (D) up to and including 1970. Mapped by the Biological Records Centre, Monkswood, UK. After Beerling *et al.* (1994).

from the parent and penetrate as deep as 2 m. Distribution of the plant in urban areas has been facilitated by movement of soil containing rhizome fragments. In riparian areas high water flows disperse fragments of plants that initiate new colonies. It is now illegal to plant knotweed in the wild; however, this is closing the gate after the horse got out. The British population is thought to be a clone from a single female individual (Beerling *et al.* 1994, Holland 2000). The lack of male plants eliminates the possibility of sexual reproduction. The historical pattern of its spread in the British Isles is shown in Figure 2.9.

The ecological theory of colonization and invasion

The idea of a plant species being introduced deliberately to a new continent, perhaps to a garden, and then 'escaping' and 'invading' habitats across a landscape, invokes a powerful image. To understand the ecological context of invasions it is useful to introduce some ecological theory that provides insight into the actual and potential impact of invading species, and on how to deal with them.

One fundamental aspect of ecological theory that is relevant to species invasions is the **species–area relationship,** first described early in the nineteenth century (Rosenzweig 1995). This describes the association between the number of species and the area sampled or available for inhabitation. The species–area relationship means that all measures of species diversity, whether of native or exotic species, will be related to the area under consideration.

Mark Williamson and Richard Fitter provide additional ecological insights into the process of invasion. They defined four stages of plant invasions as (1) introduction, (2) escape and establishment, (3) naturalization and spread, and (4) achieving pest or weed status (Figure 2.1). According to an analysis of data on British plants approximately 10% of plants pass the transitions between each of these stages and this led Williamson and Fitter (1996a) to develop what they call the 'Tens rule' for plant invasions. The likelihood of a plant species passing through these stages will depend on both the life-history characteristics of the plant and the characteristics of the environment.

These ecological insights relate introduced plants to landscape ecology – the study of how the complex spatial structure of landscapes affects ecological patterns and processes (Forman and Godron 1986, Wiens

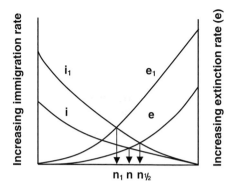

Figure 2.10. The interactions between immigration and extinction based on the theory of island biogeography. Changes in immigration rates (i) can reduce the number of species (n) if they are associated with higher extinction rates (e), or increase the number of species if the extinction rate remains unchanged. Modified from Dawson (1994).

1995, Turner *et al.* 1996, Sanderson and Harris 2000). A key early influence on landscape ecology arising from the species–area relationship was the Equilibrium Theory of Island Biogeography developed by MacArthur and Wilson (1963, 1967). MacArthur and Wilson made two empirical observations: (1) the number of species on islands is related to their area, and (2) islands become increasingly impoverished in species number with increasing distance from the nearest landmass. From this starting point, they predicted that the number of species in an island or patch of habitat is at an equilibrium between immigration and extinction. They assumed that colonization rates would vary inversely with distance from sources of colonizing species and that the number of new colonists should decline as the number of species already in the patch increases. They also assumed that the number of extinctions should also increase with the number of species present.

The introduction and invasion of exotic species is a special case of this theory. The introduction of non-native species raises the immigration rate and the effects of exotic species may increase the extinction rate, if they harm native species. Increased immigration of invasive species can therefore reduce the number of species if it causes extinction, or it can increase the number of species if the extinction rate remains unchanged as diagramed in Figure 2.10.

The Theory of Island Biogeography has been criticized recently (Brown and Lomolino 2000, Heaney 2000, Lomolino 2000), partly because it assumes the existence of equilibrium conditions. In addition, speciation was not initially considered to be a major factor in island biogeography theory because it was assumed to occur over too long a time. However, Losos and Schluter (2000) have shown that rapid speciation may contribute to species diversity on islands. Also, Whittaker (2000) recently pointed out that the colonization and ecosystem development of near-shore islands can be considered to be a special case of **succession** (see Chapter 3).

Another aspect of ecological theory related closely to MacArthur and Wilson's theory is **metapopulation biology**, where the movements of individual organisms among habitat patches are considered in terms of local extinction and recolonization (Levins 1970, Merriam 1991). Levins's key initial finding was that individuals or seeds moving in both directions between a set of patches can maintain a species in a landscape as a metapopulation, even if extinction rates in all patches are high. Others (e.g. R. Pulliam, S. Harrison) have stressed that it is much more common for population dynamics in patchy environments to be driven by movements of seeds or adults from large 'source' patches to smaller 'sink' patches. Metapopulation theory can apply to the dynamics of invasive plants when there is a balance between extinction and regeneration through seed dispersal (Box 2.8).

Landscape ecology utilizes multi-scaled approaches to look at the movement of organisms across continents, islands, landscapes and regions. It also deals with **habitat fragmentation** (the reduction of once continuous habitat to smaller and often isolated patches), and **corridors** (strips of habitat connecting patches) (Turner *et al.* 1996, Harris and Sanderson 2000). Additional factors highlighted by landscape ecologists include: **edge effects** (where some species are more likely to be found along edges or ecotones than others), and **adjacency effects** generated by the habitat type and disturbance regime in areas adjacent to a habitat of interest. Landscape ecologists also recognize that random but catastrophic disturbances such as hurricanes and floods will affect the organisms occupying habitat patches (Harris and Sanderson 2000).

Landscape ecology and invasive species

How much insight do the theories of landscape ecology and island biogeography give us into the movement of invasive species across landscapes,

Box 2.8 · *Metapopulations.*

Populations of organisms can be connected through dispersal and the sum of these becomes a metapopulation. The metapopulation structure of populations can influence their persistence in an area. If the metapopulation fluctuates regionally such that in certain years all populations decline and in others all increase, the metapopulation structure will be less effective at stabilizing population densities than would be the case if population increases and declines were out of phase (Figure 2.11). Populations of many biennial plants become temporarily extinct with the overall population persistence being a balance between extinction and regeneration through seed dispersal or from a seed bank. The metapopulation dynamics of *Senecio jacobaea* in sand dunes in the Netherlands was studied by Meijden *et al.* (1992) over 17 years (Figure 2.12). They began with 102 populations and each year scored the fate of the population. Population extinctions are greater in dry years and can be influenced by defoliation by caterpillars of cinnabar moth, *Tyria jacobaeae*. Populations in partly shaded areas serve as temporary refuges and allow reinvasion to other sites subsequently. A different pattern was seen with another biennial plant in this same system, *Cynoglossum officinale*. For this species some subpopulations declined rapidly, others more slowly and yet others increased during the study. Populations of *C. officinale* showed more asynchrony than

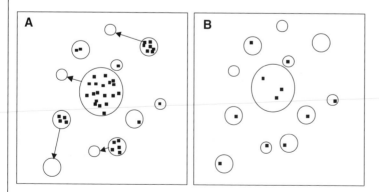

Figure 2.11. Metapopulation structure could allow source populations to reinvade habitat areas in which populations have become extinct (A), or if all populations decline at the same time, recovery of extinct populations will be slow (B).

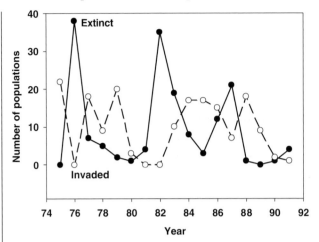

Figure 2.12. Populations of *Senecio jacobaea* in the dunes of the Netherlands become extinct and are reinvaded over time. Data from Meijden *et al.* (1992).

did *S. jacobaea* populations. Disturbance is important to maintaining these and other biennial plants.

into habitat patches and along corridors? Most data on landscape characteristics such as fragmentation and corridors come from animals, particularly birds (Dawson 1994). Turner *et al.*'s (1996) review of the loss of species from fragments of tropical rainforests, highlights the lack of studies on plants and invertebrates. Also, compared with the literature on endangered species (e.g. Arnold 1995, Harrison and Fahrig 1995), little research has examined the relationship between landscape structure and introduced species (di Castri *et al.* 1990, Brothers and Spingarn 1992, Mack *et al.* 2000). Plant invaders tend to be evaluated in terms of their impact on the ecosystem or plant community that they are invading. What makes a particular ecosystem or community susceptible to invasion is covered in Chapter 3 (e.g. Vitousek 1990, Mack 1996, Levine and D'Antonio 1999, McCann 2000, Rejmánek 2000). The studies on movement of introduced species across landscapes tend to be more qualitative (Mack 2000). However, two key questions from landscape ecology are relevant to the spread of invasive species: (1) what is the role of corridors in facilitating invasions? (2) how do landscape level factors influence invasions?

How do corridors affect the spread of introduced species?

Godron and Forman (1983) recognized four types of corridors: (1) line corridors, or narrow strips of edge habitat, (2) strip corridors, containing interior habitat, (3) stream corridors, and (4) networks, formed by the intersection of corridors. These connections between habitat fragments are generally considered to promote the movement of organisms, and thus preserve biodiversity and ecosystem and community functioning (Harris and Sanderson 2000). However, Dawson (1994) concluded that direct tests of a conduit function for corridors have revealed few convincing examples. He also concluded that 'all-purpose' corridors do not exist because each species has its own requirements for habitat, its own ability to move, and its own behaviour. Simberloff *et al.* (1992) drew similar conclusions and called for cost–benefit analyses to precede costly corridor projects. Little help is available for managers requiring practical advice on where, when and for what species corridors may act as conduits. Each case needs to be considered on its own merits. In a recent review of fragmentation studies, Debinski and Holt (2000) found that species richness and abundance relative to patch size showed inconsistent relationships, but that movement and species richness were positively affected by corridors and connectivity. Thus, there appears to be increasing evidence supporting the prediction that corridors allow species to persist in habitat fragments. However, what kinds of species are these?

Corridors may facilitate the spread of invasive species, particularly if they are associated with long edges of disturbed habitat. Roads are particularly good corridors for the movement of non-native species; they alter conditions, stress native species, and allow easier access of humans and other vectors of plant dispersal (reviewed in Trombulak and Frissell 2000). Disturbance associated with road construction provides a fertile environment for many weed species and the movement of seeds on construction equipment is an excellent dispersal mechanism. In some areas increased salinity associated with road salting has facilitated the invasion of maritime, salt-tolerant plants. Exotic species have been used for erosion control along roads and in this way road cuts serve as sources of colonizing plants capable of moving into adjacent areas (see Box 2.5B for an example). Railways have also provided routes of movement for introduced species.

The spread of exotic species is influenced by three types of barriers; biological, physical, and environmental (Figure 2.13). Parendes and Jones (2000) examined the roles of these three barriers in explaining the spatial

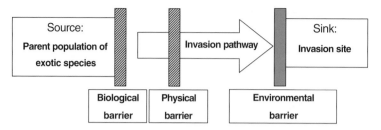

Figure 2.13. Three potential barriers to the spread of introduced plant species. After Parendes and Jones (2000).

patterns of exotic plant species along roads and streams in a forest land-scape in Oregon. They characterized habitats as being high-use roads, low-use roads, abandoned roads, and streams. Low seed production or dispersal represented a biological barrier, road use represented a physical barrier and light levels represented an environmental barrier in this study. Data were collected on the occurrence of 21 species of exotic plants in 2 m wide transects running along the edges of roads and streams. Light levels were strongly influenced by the habitat with high-use roads having the highest light levels and streams the lowest. The distribution of non-native species reflected this variation in light levels with some of the introduced plant species being more frequent in the high- and low-use road edges than abandoned roads and streams (Figure 2.14). However, there was variation in the response of these non-native species. The most frequently occurring non-native species in this study had morphological seed characteristics that suggest a high potential for wind or animal dispersal. The eight least common species had variable seed sizes and showed no clear trends in occurrence with disturbance or light levels. The authors concluded that roads and streams act as corridors or agents for dispersal, provide suitable habitat, and retain reservoirs of propagules for future invasion following disturbance for some species of exotic plants. Thus for some introduced species corridors provided by roads and streams break down all of the barriers to invasion described in Figure 2.13.

Disturbance, light levels, and connectivity all can contribute to the invasion of exotic species to new areas. This does not necessarily mean that the species will invade adjacent unsuitable habitat, such as forests, but these sites can provide reservoirs for the invasion of future disturbed sites resulting from clearcuts or flooding. The spread of exotic plants along invasion pathways warrants further investigation and must be considered in land use planning of parks and reserves.

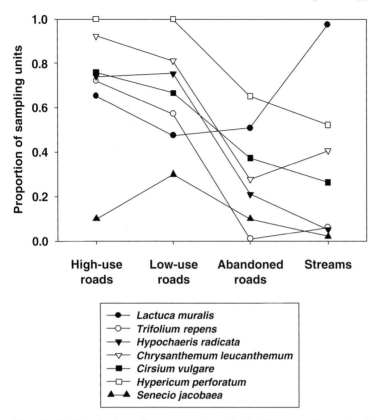

Figure 2.14. Distribution of target exotic species in 50 m sampling units along each habitat type. After Parendes and Jones (2000).

Landscape level patterns of invasion – the Lonsdale model

The basis of the island biogeography model is that the number of species will be related to island or patch size, that islands have lower species diversity than mainland sites, and that reduced levels of species introductions will reduce species diversity. We next consider how these relationships apply to non-native species.

Lonsdale (1999) developed a simple model (Box 2.9) for characterizing habitat invasibility of introduced plant species and used data from the literature to look for global patterns of invasibility at 184 sites.

To compare the success of exotic species invading different areas, it is necessary to know for all the areas how many non-indigenous species have been introduced. Of course this is not possible in most cases, as

Table 2.4. *Relationships of habitat factors to factors determining E, the number of exotic species in an area*

Habitat factors	Impacts on invasives
Disturbance	High disturbance – high S_v
Recovery from disturbance	High recovery – low S_v
Resistance of native species	High resistance – low S_v
Ecosystem resistance	High resistance – low S_v or low S_h
Invasibility	Higher invasibility – high S
Invasion potential of exotic species	High potential – higher S_v, S_h, or S_c
Propagule pressure – number arriving	Propagule pressure causes I to vary

S = survival of species, I = number of introduced species.
After Lonsdale (1999).

Box 2.9 · *The Lonsdale model of invasibility.*

E = IS
(Exotic species) = (Species introduced) (Survival in new range)

$I = I_a + I_i$
(Species introduced) = (Introduced accidentally)
                                    + (Introduced intentionally)

$S = S_v\ S_h\ S_c\ S_m$
(Survival) = (Survival after competition) (Survival after herbivory
                    and pathogen) (Survival after chance events) (Survival
                    after maladaptation)

---

there will be no record of most unsuccessful introductions. Therefore the estimates of invasion or 'I' in Lonsdale's model (see Box 2.9) will always be conservative. The resistance of ecosystems to invasion will be determined by the factors that influence the survival and success of the propagules. Some of these factors are outlined in Table 2.4 with their impacts on the terms in Box 2.9.

In addition to these relationships, the number of native species and the number of introduced species will often be influenced by the size of the area inhabited (Figure 2.15). Therefore patterns of the invasion of non-native species generally fit the species–area relationship on which island

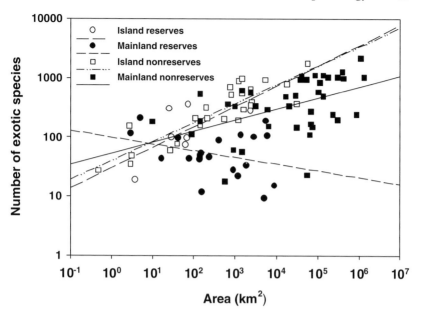

Figure 2.15. The relationship between the number of exotic species and site area for 104 sites around the world indicating island reserves, island non-reserves, mainland reserves, and mainland non-reserves. Lines are fitted from regressions for each data set and differ from those of Lonsdale (1999), who used a different model.

biogeography theory is based, although this pattern was not apparent for mainland reserves in Figure 2.15. However, because the relationship between area and native species has a steeper slope than that relating area to the number of exotic species (particularly for mainland reserves), the fraction of exotic species actually declines with area (Figure 2.16). This means that any comparison of invasibility among sites has to be scaled for size. One possible way to scale the data for the interaction of area and species is to use the percentage of exotic species out of all species at the site. This is the metric that Lonsdale used in his comparisons to look for patterns in invasibility and to test some hypotheses from the invasion literature.

Data on the fraction of exotic species on islands and mainland sites show a very different pattern from that predicted by island biogeography. Islands have more exotic species than mainlands (Figure 2.15, 17 open symbols). The explanation for this may be related to the low resistance of island ecosystems to invasions, or to the high levels of disturbance there (Table 2.5).

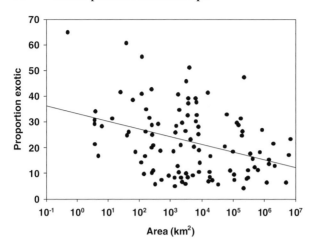

Figure 2.16. The relationship between the fraction of the flora that is exotic and the area of the site. $R = 0.35$, $n = 104$. After Lonsdale (1999).

An encouraging result of Lonsdale's analysis is that for the mainland sites, reserves had fewer exotic species than non-reserves (Figure 2.17). Therefore reserves help to preserve natural biodiversity. The biome with the greatest fraction of exotic plant species was temperate agricultural and urban areas. Savannas, wet tropics and deserts had the least. Other comparisons of the invasibility of areas must take these associations into consideration. Overall the number of exotic species in an area was strongly, positively related to the number of native species and to the size of the area. Together, area and the number of native species explained 70% of the variation in exotic species richness. Further analysis by Lonsdale indicated that the New World is more invaded than the Old World, and that the richness of exotic species increases with the richness of native species, but that there was little difference between tropical and temperate areas (Table 2.5).

Lonsdale's analysis supports MacDonald *et al.*'s (1989) conclusions that the number of exotics in reserves is related to the number of visitors after correcting for reserve size and native species diversity. Various alternative explanations are possible for these relationships (Table 2.5). However, the human visitation could be taken as a correlate for isolation or distance from a mainland (as envisioned by MacArthur and Wilson). While only experimentation will reveal the mechanisms influencing the invasibility of areas, the analyses of Lonsdale and others have provided the essential ground work for further studies.

Table 2.5. *Summary of generalizations about invasions, their original explanations, results of analysis and alternative explanations*

Original hypothesis	Explanation	Result	Alternative explanations
New World more invasible than the Old World	Old World species have higher invasion potential	No consistent pattern unless native diversity is factored out	Immigration rates to the New World greater. (Higher I in Lonsdale's model)
Rich communities are less invasible	Rich communities are more stable and have fewer empty niches	Opposite is true – more diverse communities have more exotic species	Richer plant communities indicate greater habitat diversity and more opportunities for invasion
Temperate ecosystems are more invasible than tropical ecosystems	Temperate ecosystems are less species rich	Consistent pattern but this contradicts the positive association of diversity and invasion at the community level	Species from temperate ecosystems: 1 better invaders 2 natives less resistant 3 seed immigration higher
Temperate islands are less invaded than tropical islands	None offered	No significant effect	Not necessary
Islands are more invaded than mainland areas	Islands are species poor	Consistent with generalization but not the explanation	Island ecosystems are: 1 less resistant 2 more disturbed 3 invaders more invasive 4 natives less resistant to disturbance 5 seed immigration higher
Increases in human visitation increases exotic species richness	Visitors increase propagule pressure (I in the Lonsdale model)	Consistent	Visitors increase disturbance and reduce isolation

Based on Lonsdale (1999).

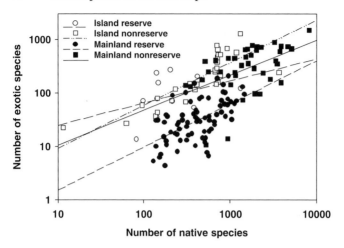

Figure 2.17. The relationship between the number of exotic plant species and the number of native plant species for 177 sites and regions broken down into island reserves, island non-reserves, mainland reserves, and mainland non-reserves. After Lonsdale (1999).

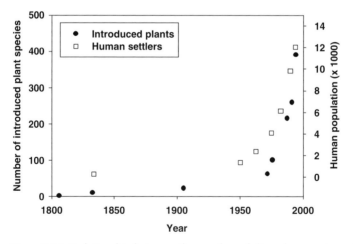

Figure 2.18. Relationship between the number of alien plant species and permanent human settlers in Galápagos. After Mauchamp (1997).

The relationship between the human population and the number of alien species in the Galápagos Islands clearly reveals that humans are a driving force for introducing plant species on both local and larger scales (Figure 2.18). However, human visitors are not always a factor in spreading exotic species to reserves. Turner and Corlett (1996) studied Bukit

Timah Nature Reserve in Singapore, a 164 ha fragment of primary low-land rainforest that was cut off from other forests in 1985 when an eight-lane highway was built. Ten years later this area still retained its interior forest character. Only one introduced plant species, the melastomataceous shrub, *Clidemia hirta*, had widely invaded the forest, despite there being many thousands of human visitors each year and an extensive network of trails (Turner and Corlett 1996). Clearly, factors other than human visits, such as deliberate dispersal of non-native species or the availability of introduced plants in surrounding areas, can be important.

Finally, edge theory predicts that more non-native species will be found near habitat edges. In the case of forest habitats, increased amounts of forest edge and increased levels of habitat disturbance should favor colonization by introduced species (Harris and Sanderson 2000). Brothers and Spingarn (1992) found this to be so. In islands of old growth, upland mesic or flatwoods forest in Indiana the species richness and frequency of non-native species declined significantly with distance from the forest edge. The decline was attributed to reduced light levels in the forest interior. Limited dispersal and the lack of disturbance in the interior areas of the forests probably also influenced the spread of invasive species. Interestingly, non-native species were more common on the southern edges of forests compared with northern edges. The most successful invaders were mostly escaped ornamentals.

A case study of habitat fragmentation, land use and the numbers of introduced plant species is described in Box 2.10. This study demonstrates

---

Box 2.10 · *A case study of introduced plants in fragmented forests.*

The rural–urban fringe of Toronto, in southern Ontario, Canada, is one of the fastest growing human populations in Canada. Fringe areas contain a mixture of fragmented agricultural and forested land, and in the last few decades the landscape structure has changed considerably. Increasing forest edge associated with fragmentation would be expected to favor introduced plant species (Hobbs 1989). Elizabeth Billyard (1996) studied these fragmented forests to ask two questions:

1 how does landscape structure affect the likelihood of finding an introduced species in a particular place, and
2 what are the landscape-level characteristics of invasible habitats, which are those with the most introduced species?

Billyard tested the hypothesis that the number and frequency of occurrence of introduced plant species will be related to (1) the edge aspect of a forest patch – whether north- or south-facing (Brothers and Spingarn 1992), (2) the surrounding land use (Juxtaposition theory), (3) woodlot area, (4) interior forest (percentage of total forest area), and (5) disturbance defined as the percentage decrease in forest area over a specified period.

To identify landscape level characteristics aerial photographs of the region were used to identify patches of upland forest at least 1 ha in size in 1992. These patches were also older than 50 years and in mid- to late-successional stages. Patches present on the photographs in both 1971 and 1992 were selected for study. The percentage change in forest fragment size from 1971 to 1992 was calculated. In June and July 1995, 19 forest fragments were sampled for the presence of introduced plant species, and the surrounding land use was identified. The ecotone between the forest edge and the surrounding landscape was surveyed. Sampling transects 4 m wide, 50 m long and parallel to the forest edge were established on both the north and south aspects of the woodlot. Five of transects were run, at 2 m outside the edge, and 2 m, 8 m, 20 m and 50 m inside. All herbaceous species in each transect were identified.

Forest in this region was highly fragmented. From 1971 to 1992, the number of woodlots decreased by 15%. Mostly smaller lots were lost, so that the median woodlot size actually rose (3.56 ha in 1971 and 3.86 ha in 1992). There was also a 34% reduction in the number of hedgerows resulting in decreased connectivity between the woodlots. On average, the 19 woodlots surveyed had decreased in size by 25%.

The number of introduced species declined with increasing distance from forest edge (Figure 2.19). A multiple regression showed that the only significant landscape level relationship was between the number of introduced plant species and disturbance, which was defined as the proportion of woodlot area lost from 1971 to 1992. Other factors such as woodlot area and percentage of interior forest were not significantly correlated with the number of introduced species in a woodlot. Thus, the importance of edges emerged.

The number of introduced plant species varied significantly according to surrounding land use and was greatest when the adjacent area was residential, followed by road, field and cropland (Figure 2.20). It

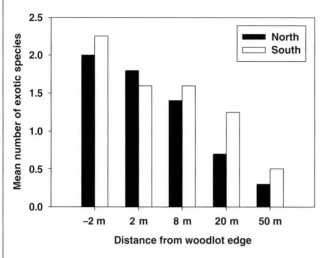

Figure 2.19. Number of exotic herbs on transects at different distances from woodlot edges (−2 m is outside) for study woodlots in King township, Ontario. Number of transects is 38. From Billyard (2001).

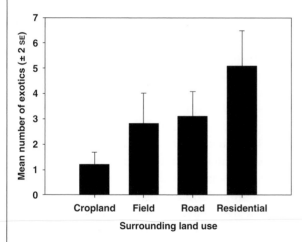

Figure 2.20. The relationship between the mean number of exotic herbs and surrounding landuse for study woodlots in King township, Ontario. From Billyard (2001).

is likely that woodlots bordered by a residential area experienced higher levels of human use and visitation, for example by people walking dogs. In this way, dispersal of introduced species was assisted.

the usefulness of aerial photography for characterizing habitat fragmentation and changes in patch sizes, when it is combined with field observations of the prevalence of native and exotic plant species. As with the forests in Indiana introduced species were more common on the southern edges of forest patches (Figure 2.19) and disturbance measured as habitat loss and surrounding land use was a good predictor of invasion by introduced plants.

Conclusions

Introduced plants are a major component of ecosystems, particularly in those parts of the world dominated by human invasions. They often make up a quarter of plant species and, in some places, even more of the total biota. Humans have both intentionally and accidentally increased the immigration rates of plant species and a relatively small, but important, percentage of these introduced species have become serious environmental and agricultural weeds. While many introduced animals have clearly caused extinctions of native species, introduced plants have not been shown to have this impact. However, the dominance of invaders in some habitats must influence the density of the species that were previously common there and they very likely change the suitability of the habitat for other organisms. Terminology related to introduced species is varied and places undue stress on the negative impacts of some species. This has caused some to question if concerns about non-indigenous species are justified or are an expression of a fear of change. Some people question how much plants should be valued for their diversity, beauty and geographic history or for their native status.

The flow of new species around the world can be viewed in the light of island biogeography and metapopulation theory, and in the context of landscape ecology. With the importation of species associated with the global movement of humans and international trade in plants, distance from the source is no longer a major issue. In conservation biology, the creation of corridors is often promoted to help animals move between patches. These same corridors, however, can facilitate the spread of exotic species into habitat reserves.

3 · Biological invasions in the context of plant communities

In the first part of this chapter, we consider what makes an ecosystem or community prone to invasion by an introduced species. In the second part we look at the impacts of invasive plant species on communities and ecosystems. These two issues are critical to the interpretation that non-indigenous species are large contributors to the current biodiversity crisis (D'Antonio and Vitousek 1992, Chapin *et al.* 1998). To begin we briefly review relevant theory on the role of plants in communities in a broader ecological framework. In particular we begin by discussing the invasion of plants into communities as a dimension of the succession process.

Part 1 – Characteristics of native plant communities that influence plant invasions

The invasion of an introduced species into communities is a special case of plant succession. Succession, which was a major focus in early studies of plant ecology, determines the diversity of plant species in communities and this process continues to fascinate ecologists. In the late 1800s, Eugene Warming wrote the first book on plant ecology and in it he recognized that abiotic conditions such as soil type and soil moisture influence plant distributions (Sheail 1987).

Frederick Clements (1916) developed the first theory of plant succession and described the stages of vegetation development (Bradshaw 1993, Miles and Walton 1993). Clements viewed plant community development as a process that begins with bare substrate and progresses through a series of 'seres' or stages. Each stage gives way to the next until the 'climax' community is reached. He viewed the plant community as a super-organism, with each stage being comprised of a complementary and fixed association of plant species. This idea persists and plant communities are today still described by their dominant species associations. However, Clements' views conflicted with those of H. A. Gleason (1926), who considered that plant communities change over time in a stochastic manner with plant species arriving by chance and, depending on prevailing conditions, becoming established.

The colonization of bare substrate by plants, and the subsequent development of soil, are known as primary or autogenic succession (Chapin 1993, Walker 1993). This differs from secondary succession in which vegetation is destroyed for some reason leaving an existing seed bank and developed soil, as well as the opportunity for plants to immigrate from nearby habitats. While the overall processes involved in primary and secondary succession are similar, differences exist between them, namely, the unstable substrate and the lack of both nitrogen reserves and pre-existing propagules in the case of primary succession (Miles and Walton 1993).

In contrast to Clements' notion of the plant community as a super-organism, Whittaker (1970) proposed the 'continuum concept' based on his idea that each species occurs along its own abiotic gradient. Particular plant associations are apparent when some similarly adapted species co-occur along these gradients. Thus even 'climax' communities are highly dynamic. The successional sequence is frequently interrupted by periodic disturbances. For example, in forests trees blow down and cause gaps and in grasslands fires change the species composition.

The processes by which new plant species enter a successional sequence are variable. Connell and Slatyer (1977) summarized three different mechanisms that can underlie changes in plant communities. The first, **facilitation**, involves the early colonizers altering abiotic conditions and making them more suitable for establishment and growth of later arrivals. This is the 'classical' view of succession. In the **tolerance** model of succession, later-arriving species are better able to outcompete earlier colonizers. In the **inhibition** model, the early colonizers resist invasion of later species, and these can only become established after the original species are removed by disturbance.

Plant communities are highly dynamic, but predicting the pathways or trajectories of change is difficult (e.g. Morrison and Yarranton 1974, Yarranton and Morrison 1974). Michael Rosenzweig (1995) remarked that 'in the first few years of succession, plant diversity rises. But I cannot say much more than that!' Whether succession is really a single process remains controversial (Crawley 1997b). However, conservation and restoration involve managing towards an 'endpoint' which is often defined in North America as a state that existed prior to the disturbance caused by European colonization. This primary focus on achieving a specific assemblage or group of plant species tends to obscure the fact that

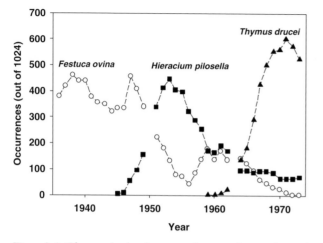

Figure 3.1. Changes in dominant species over time as shown by the occurrences of *Festuca ovina*, *Hieracium pilosella* and *Thymus drucei* in 1024 plots (1.27 × 1.27 cm) following the removal of grazing by rabbits. After Watt (1981b).

plant communities are dynamic, and change continually with or without human interference.

Detailed, long-term data from permanent plots allow the evaluation of natural variation in community composition, but relatively few such datasets exist. Where they do (e.g. Watt 1981a,b, Milchunas and Lauenroth 1995, Klotzli and Grootjans 2001) they show that plant populations and communities vary enormously over decades, and not always as predicted in response to perturbations such as grazing and abiotic alterations. For example, annual observations of a fenced plot in England show that switches in the occurrences of dominant species (Figure 3.1) are related to winter frost and spring drought (Watt 1981b). In the following sections, we consider several processes that affect the likelihood of new species colonizing a plant community, including introduced and invasive plant species.

Disturbance and succession

The notion that plant communities reach an equilibrium is at the heart of Clements' theory of succession and MacArthur and Wilson's (1967) theory of island biogeography (see Chapter 2). The equilibrium concept has persisted (Duchesne 1994), but it is increasingly regarded as flawed (Lomolino 2000). Disturbance can cause a plant community to revert to

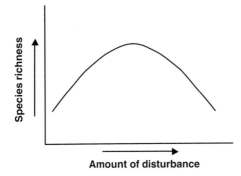

Figure 3.2. Simple diagram of the intermediate disturbance hypothesis that predicts the highest species diversity at intermediate levels of disturbance.

an earlier successional stage (Crawley 1983) and allow colonization by different plant species (Canham and Loucks 1984).

Changes to the species diversity of plant communities need not be directional. The intermediate disturbance hypothesis explains variation in species diversity in terms of the intensity and frequency of disturbance. Species diversity is predicted to be highest at intermediate levels of disturbance (Figure 3.2) (Petraitis *et al.* 1989).

An experimental test of the intermediate disturbance hypothesis was done in the Konza Prairie, in Kansas. Experimental plots were burned at 1, 2, 4, and 20 year intervals. The results did not support the intermediate disturbance hypothesis; plots burned annually had the fewest species and those burned every 20 years had the most (Figure 3.3A). No optimum level of disturbance was obvious. However, for plots that were studied following fire treatment, the species richness was highest at an intermediate time (Figure 3.3B). Collins *et al.* (1995) interpret these results to indicate that fire causes extinctions of species, but recovery from fire is a balance between immigration and extinction. Richness of both native and introduced species was the highest in unburned plots. Overall, experimental tests of the intermediate disturbance hypothesis have had contradictory results (reviewed in Begon *et al.* 1996). Defining the specific intermediate level of disturbance that could promote diverse communities may be difficult.

Grime's C–S–R model of succession

The life-history strategies of plant species can also influence the development of plant communities. Grime (1974, 1979) proposed that the

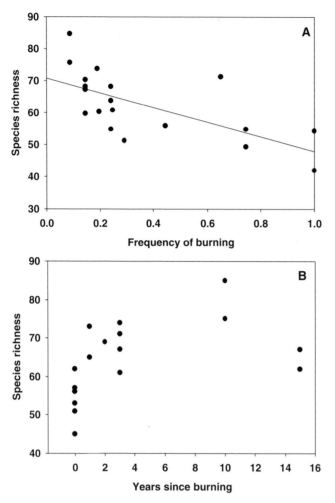

Figure 3.3. Species richness (number of species), and (A) frequency of fires (number of times burned), (B) years since burning for experimental plots of tall grass prairie in Kansas. From Collins *et al.* (1995).

succession of plant communities is determined by the composition of plant functional groups. He identified three types of plant species; C-types that are good competitors, S-types that are well adapted to stressful conditions such as low soil moisture and nutrients, and R-types, ruderals that cope well in disturbed environments. Ruderals are generally small annuals with rapid growth and high seed production. Grime places plant species in a competition, stress, disturbance matrix (Figure 3.4). Plants with different

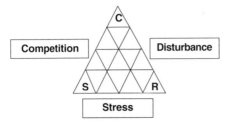

Figure 3.4. Model of the C–S–R theory. Plant attributes evolve depending on the strength of competition (C), stress (S) and disturbance (ruderal) (R) characteristics. Modified from Grime (1979).

characteristics are predicted to be more or less common during different stages of succession. For example, ruderals dominate in early succession. At later stages, the competitors and stress tolerators will be more common, although if a disturbance occurs, then ruderal species will reappear. This 'functional group' approach has gained support in recent years, due both to tests with multivariate statistic models (Grime *et al.* 1997), and to its adoption by more plant ecologists (Gitay and Noble 1997).

Disturbance and the invasion of plant species

Disturbance is considered to promote the colonization (establishment) and expansion of non-native species (Elton 1958, di Castri 1989) and intact plant communities should be resistant to invasion. Thompson (1994) used the C–S–R classification to determine whether human-induced disturbance increases and improves conditions for introduced plants. He looked at changes in the relative abundance of the different functional groups of plant species in the flora of various European countries (Republic of Ireland, Scotland, Northern Ireland, Wales, The Netherlands and England) over a period of just under 50 years, and he related this to human population density. Plants were grouped by their life-history characteristics and given values associated with their location on Grime's stress/disturbance matrix (Figure 3.4). In addition plant species were scored for other attributes: canopy height, life history, seed bank, wind dispersal, clonal growth, dispersule weight, flowering period, and lateral spread. Whether the species had decreased or increased in abundance over time was then determined.

The main plant characteristic that explained the change in species abundance relative to human population density, was the 'S-score', the measure of stress tolerance of plant species. Plant species with a high

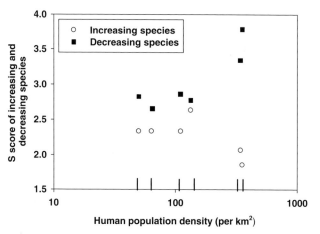

Figure 3.5. The relationship between stress tolerance (S-score) of increasing and declining plant species and human population density in six European countries, indicated on the axis by vertical lines. From left to right countries are Scotland, the Republic of Ireland, Northern Ireland, Wales, The Netherlands, and England. S-scores for increasing and declining plant species are significantly different for England and The Netherlands ($p < 0.001$). After Thompson (1994).

S-score tend to grow more slowly than those with a low S-score. Thompson (1994) found that in the two countries with the highest human population densities, The Netherlands and England, species that had increased abundances were those with a lower S-score (fast growers) (Figure 3.5). In contrast, the species with declining abundances at these high human population densities, were those with a high S-score (slow growers). In the other countries species that increased or decreased in frequency did not differ significantly. Interestingly, life-history characteristics such as seed bank, wind dispersal, flowering period and seed weight made little or no contribution to the discriminant function used to categorize plant species.

To further evaluate the relationship of disturbance to invasion, Lozon and MacIsaac (1997) reviewed 63 studies of 299 introduced plant species. Although the importance of disturbance varied, the establishment of 86% of the exotic plant species studied was associated with disturbance. Of the 257 plant species for which disturbance promoted establishment, 68% (174) were reported also to require disturbance for range expansion. In a second survey, these authors considered the type of disturbance that facilitated establishment of 404 introduced plant species. Human-induced disturbance was associated with 97% of the established exotic plant

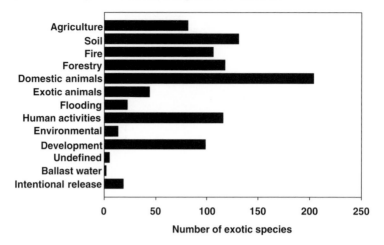

Figure 3.6. Activities associated with the establishment of introduced plant species. After Lozon and MacIsaac (1997).

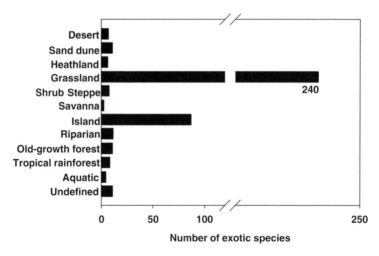

Figure 3.7. Habitats in which disturbance facilitated the establishment of introduced plant species. Old growth forests are in temperate regions. After Lozon and MacIsaac (1997).

species. These involved a number of different factors including agriculture and grazing animals (Figure 3.6). Most plant introductions and expansions were found to be on islands and in grassland habitats, many of which are likely to have been transformed by agricultural activities (Figure 3.7). These results suggest that many invasive plants are ruderals as defined by Grime.

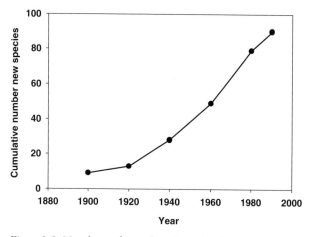

Figure 3.8. Numbers of introduced species documented in Israel, 1927–1990. Data from Dafni and Heller (1990).

How disturbance actually facilitates plant establishment is less well known. The creation of open patches for seedling establishment (see Chapter 4) may be the most important factor. Disturbances such as fire, flooding, logging or even ploughing could also increase the available mineral resources and sun exposure. Selective grazing could reduce the competitiveness of palatable, native plants and facilitate the establishment of less palatable exotic species. The strong association of disturbance with plant establishment suggests that dispersal is not nearly as strong a limiting factor as the availability of sites for germination. However, Tilman (1997) argued that limited recruitment did influence the diversity of grasslands. In his experiments the addition of seeds from exotic plant species increased diversity in experimental plots regardless of the initial diversity of the plot. Clearly the seeds must first reach the new sites in order to become established.

Disturbance and recruitment of introduced species are both associated with human activities. Dafni and Heller (1990) proposed that the transition from extensive to intensive agriculture in Israel following independence in 1948, combined with the increase in the human population, caused the increase in numbers of introduced species at that time (Figure 3.8).

Di Castri (1989) pointed out that it is important to distinguish between long-term disturbance and the disturbance at the time of the invasion. In some cases humans may suppress the frequency of disturbance and this could influence the invasibility or species composition of communities. Two kinds of disturbances that are frequently altered are fire and flooding.

Whether fire as a disturbance facilitates invasions by introduced species is unresolved. Many prairie and savanna habitats are recognized as being fire-dependent communities that are maintained by burning. When fire is suppressed, woody shrubs and trees can invade (Crawley 1983). Extensive experiments in the French Mediterranean region, with prescribed burns carried out at different frequencies, did not result in the long-term increase of introduced species (Trabaud 1990). Similar results were found in native grasslands in Kansas, USA (Smith and Knapp 1999), where the cover of introduced C3 grasses was significantly lower in annually burned watersheds compared with unburned watersheds. The number of both native and exotic species was higher in unburned plots. Thus, disturbance associated with fire suppressed or controlled introduced species in these studies. There is one well-documented instance of an invasive grass, cheatgrass, *Bromus tectorum*, becoming dominant and increasing the frequency of the fire cycle in the western USA, but it is not clear whether an initial alteration in the natural fire cycle enabled cheatgrass to colonize (D'Antonio and Vitousek 1992). Collins (2000) found that burned tallgrass prairie communities changed more than unburned ones, but he did not relate this to the proportions of introduced species.

Herbivory and introduced plant species

Vertebrate and invertebrate herbivores can impact plant community composition (Crawley 1983). Unlike animals that are killed and eaten by predators, an established plant that is eaten by a herbivore may not die, although its growth and reproduction will be reduced. The question of whether herbivores control plant populations or if herbivores are controlled by the availability of their forage species has been long debated (see Chapter 5). Herbivores may disturb plant communities such that plant species richness responds along a herbivory gradient in a pattern similar to that shown by physical disturbances in the intermediate disturbance hypothesis (Figure 3.2). Large vertebrate herbivores both directly remove plants or plant tissue and indirectly effect the ecosystem through trampling and fertilization (Harper 1977). At low levels of grazing pressure, one or two plant species may outcompete others (Crawley 1983).

Charles Darwin (1859) was the first to suggest that plant species can invade native communities because they are avoided by herbivores in their new habitats. This interpretation was promoted a century later by

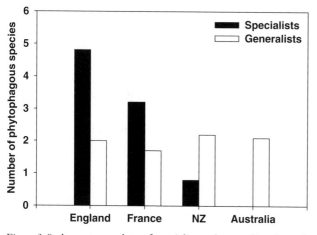

Figure 3.9. Average number of specialist and generalist phytophage species per broom bush in England, France, New Zealand (NZ), and Australia. After Memmott *et al.* (2000).

Elton (1958) and is still widely held. An underlying assumption is that in native habitats, specialist herbivores suppress the population densities of their food plants (see Chapter 7). Darwin continued this argument by showing that, according to information in Asa Gray's *Manual of the Flora of the Northern United States*, introduced plant species tend to represent new genera. Of 260 introduced plant species in 162 genera, 100 genera were new to North America (Rejmánek 1996). Similarly, 112 European grasses have become naturalized in California and, of these, 43 belong to genera native to California while 69 species belong to 39 genera that are new to California. Plants without close relatives are likely to lack specialist herbivores in the area of introduction. One study that actually quantified the invertebrate fauna on an introduced plant, Scotch broom (*Cytisus scoparius*), in its native and introduced ranges found that the total number of invertebrate species attacking plants was lower in the areas of introduction, but the number of generalist herbivores was the same in native and introduced areas (Memmott *et al.* 2000) (Figure 3.9). Broom is taxonomically distinct from other plant species in Australia and New Zealand where it was introduced and to which comparisons were made. This example quantifies the speculation that introduced plants will lack specialist herbivores in exotic habitats.

The relationship between the taxonomic relatedness of native and introduced plant species will be discussed further in Chapter 7 on biological control. Nearly all of the cases of non-target impacts by specialist insect herbivores introduced as biological control agents are on closely related indigenous plants (Pemberton 2000b). Also to be discussed in Chapter 7 is the underlying assumption that specialist herbivores, insects, and diseases reduce the densities of their host plants. It seems that a lack of specialist herbivores may facilitate plant invasions, but that other factors such as soil microbes might also influence invasiveness (see Chapter 6).

Influences of generalist and specialist herbivores on community invasibility

Maron and Vilà (2001) reviewed evidence for the natural enemies hypothesis, i.e. that introduced plants do well because they lack specialist natural enemies, and for what they called the biotic resistance hypothesis, i.e. that introduced species are superior competitors because native herbivores apply little biotic resistance to the invasion of plants. They conclude that despite considerable work on plant–herbivore interactions, there remains little understanding of the impact of herbivores on the distribution, abundance, and population dynamics of plants. Even though introduced plants accumulate a large number of generalist insect herbivores in their new habitat (114 insect species on *Mimosa pigra* in Australia, over 80 species on introduced thistles in Canada, and 30–47 species on thistles in California) (see also Figure 3.9) these tend to have very little impact on the densities of the introduced plant species.

In a review of 18 studies, Maron and Vilà (2001) found that on average native herbivores reduced the performance of seeds and seedlings of introduced plants by a third, they reduced the performance of adult exotic plants by one half, and attack by herbivores on average killed over half the plants. Therefore, although herbivore attack on introduced plants is relatively high, it is not sufficient to suppress the invasion of some plant species. Maron and Vilà recommend more comparative studies to determine the ecological conditions and life-history attributes that predict if herbivores will influence plant population dynamics.

Resistance of invasive species to grazing

In grassland systems with large mammal grazing, invasive species are commonly considered to be superior competitors because of their evolved resistance to grazing. For example, Richards (1984) reported that the introduced grass, *Agropyron desertorum* is more tolerant to heavy grazing than is the native species, *A. spicatum* in western USA. Many invasive plants are considered to be weeds because they are toxic to grazing animals and therefore avoided. Tansy ragwort, *Senecio jacobaea*, St John's wort, *Hypericum perforatum*, and blackberries, *Rubus* spp., are just a few examples of successful invasive species that are defended against herbivores. If this is a common association we might expect non-native species to be more likely to invade plant communities in the presence of vertebrate herbivores.

Studies of herbivore removal show that it is not always easy to predict how plant communities will respond to invasion under different levels of grazing. Zeevalking and Fresco (1977) studied plant diversity in sand dunes under different intensities of grazing by rabbits. Their results fit the intermediate disturbance hypothesis. Under intense disturbance, there was an overall decline in plant species richness. Reduced herbivory should therefore increase species richness by releasing the plant species present but formerly suppressed. Where there are few or no non-native plants, the native flora may benefit from reduced hervivory, but introduced plant species may also have been suppressed by herbivory and these too could resurge with reduced pressure (see Chapter 2). In the study of fire disturbance in native grasslands in Kansas cited above (Smith and Knapp 1999), grazing by bison, *Bos bison*, increased the species richness of both native and exotic grasses in annually burned sites.

A spectacular example of control of a non-native plant species by non-native herbivores, feral pigs and goats, was observed in Hawaii. Removal of pigs and goats led to an explosion in growth of a previously unobserved non-native vine, *Operculina ventricosa*, that subsequently blanketed the hillsides (Zavaleta *et al.* 2001).

In a unique study of the diversity of native and exotic species in 26 long-term grazing exclosures in Colorado, Wyoming, Montana, and South Dakota, Stohlgren *et al.* (1999) rejected the hypothesis that large mammal grazing increases the diversity of exotic species. They found nearly identical diversity of native and exotic species in grazed and un-grazed, 1000 m^2 plots. Grazing pressure did not obviously facilitate the invasion of exotic species in Rocky Mountain grasslands although two exotic grass species showed different patterns. The sod forming Kentucky blue grass, *Poa pratensis*, had a consistent pattern of higher cover in ungrazed plots while the annual grass *Bromus japonicus* was consistently higher in sites grazed by a variety of grazers including bison, elk, moose, deer, cattle, sheep, and horses.

Stohlgren *et al.*'s finding contrasts with results from other studies (e.g. Pettit *et al.* 1998, Silori and Mishra 2001). In Australia, the cover of

Table 3.1. *Richness of species of native and exotic species in long-term grazing experiments in Rocky Mountain grasslands*

Plot type	Native species richness	Exotic species richness
Ungrazed	31.5 ± 2.5	3.1 ± 0.5
Adjacent grazed	32.6 ± 2.8	3.2 ± 0.6
Randomly selected, grazed	31.6 ± 2.9	3.2 ± 0.6

Data from Stohlgren *et al.* (1999).

non-native annual grasses and herbs increased significantly in heavily-grazed compared with lightly-grazed sites (Pettit *et al.* 1998). The introduced species were grasses and the native species were perennial shrubs that were vulnerable to grazing. In southern India, cattle-grazing adjacent to the Mudumalai Wildlife Sanctuary reduced the recruitment of native trees and increased the cover of introduced species such as *Lantana camara* (Silori and Mishra 2001).

Our review of these several studies suggests that the impacts of grazing mammals vary with the type of vegetation, the type of grazing mammals, and the levels of other types of disturbance. Clearly grazing animals can have enormous impacts on the structure of plant communities, but their particular roles in promoting plant species invasion must be evaluated experimentally on a case-by-case basis.

Interspecific competition and plant invasion

During succession clear shifts in plant communities take place, and interspecific competition can be a driving force in these changes (e.g. Grace and Tilman 1990). An experimental approach to examining the competitive abilities of introduced species is to examine the response of native species to the removal of the introduced species. Brooks (2000) showed that thinning the seedlings of three introduced grass species, *Bromus madritensis* ssp. *rubens*, *Schismus arabicus* and *S. barbatus*, increased density and biomass of native annuals at three sites in the Mojave Desert, USA, in only one year of a two-year study. This was a year of high productivity and species richness (Figure 3.10). Thus, in some years, the introduced grasses outcompeted the native annuals, but not always. This is an area of research that requires more attention.

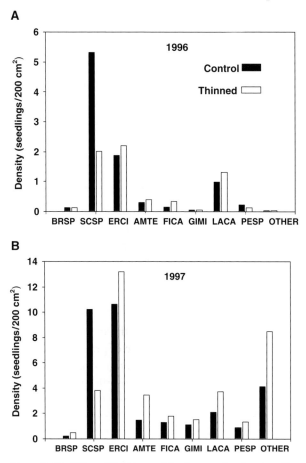

Figure 3.10. Effects of *Schismus* thinning on the density of annual plants beneath the canopy in (A) 1996, (B) 1997. Values are averages of 25 replicates at three study sites in the Mojave Desert. Species are *Bromus* sp. BRSP, *Schismus* sp. SCSP, *Erodium cicutarium* ERCI, *Amsinckia tesselata* AMTE, *Filago californica* FICA, *Gilia minor* GIMI, *Lasthenia californica* LACA, and *Pectocarya* sp. PESP. After Brooks (2000).

Since the mid–1800s many have observed that introduced plants appear to be larger and more vigorous than native species (reviewed in Thébaud and Simberloff 2001). Crawley (1987) tested this perception by comparing recorded heights of European plants growing in California to those same species in Europe. Forty-three percent of 228 European plant species were larger in California than in Europe and 29% were smaller in California than in Europe. Crawley *et al.* (1996) subsequently compared the growth of native and non-native plants in Britain and concluded that the non-native plants grew larger and produced larger seeds. Similarly, Willis and Blossey (1999) used common garden experiments to compare the growth of purple loosestrife from indigenous and non-indigenous populations and found that North American populations grew better regardless of where they were grown. Willis *et al.* (2000) carried out additional common garden experiments for four other invasive weeds and found little evidence for alien species to be more productive than natives. Most recently, Thébaud and Simberloff (2001) compared the recorded maximum heights of European plant species growing in Europe, in California and in the Carolinas, as well as California and Carolina plants growing in Europe and North America. Californian and Carolinan species are taller than European species in both North America and Europe (Figure 3.11). However, patterns do not hold across the board and plants in the Poaceae are not taller in either area and European species of Asteraceae are taller in California but Californian species are not taller in Europe (Thébaud and Simberloff 2001). Clearly patterns of plant growth and size are variable between native and exotic habitats and responses vary among species and populations. In some cases the greater vigor of introduced plants could be related to their ability to invade.

Observations of plants being larger in exotic habitats led Blossey and Nötzold (1995) to hypothesize that plants expended less energy on anti-herbivore defenses when they lacked specialist herbivores and therefore could devote more resources to growth. In a limited experiment Blossey and Nötzold (1995) evaluated the herbivore defenses of purple loosestrife, *Lythrum salicaria*, of European and North American origin with two insect species. The success of a leaf feeding beetle did not differ between plants from the two sources, but survival and growth of a root feeding beetle was better when reared on North American plants suggesting that they were less well defended. These tests were expanded by Willis *et al.* (1999) who compared plant quality for insect growth and fitness of plants from six indigenous and six non-indigenous populations of loosestrife. No significant differences occurred in leaf beetle fecundity, mass, or size when grown on plants from the different plant populations with the exception of one indigenous plant population on which no beetles survived.

One must conclude from these analyses that no consistent patterns in plant sizes between foreign and native habitats exist and evidence does not support the hypothesis that introduced plants devote fewer resources to herbivore defenses.

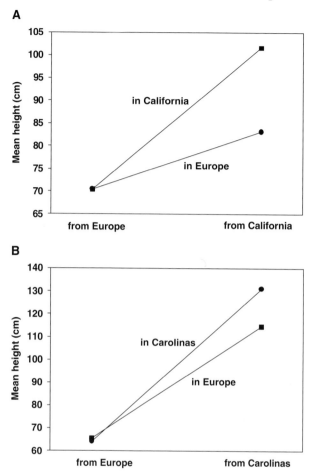

Figure 3.11. Mean plant height as a function of geographic origin and current location. (A) Europe versus California, (B) Europe versus the Carolinas. Values are back-transformed least square means. After Thébaud and Simberloff (2001).

Chromosome doubling is a common evolutionary event in plants and plants with higher numbers of duplicate chromosomes are often more successful. Sulphur cinquefoil, *Potentilla recta*, which invaded North America over 100 years ago has varying levels of ploidy. Seedling establishment is critical to the population dynamics of this species and varies between years. A comparison of North American, central European and eastern European populations showed that although hexaploids had larger seeds and higher initial growth rate, tetraploids grew faster after five weeks post-germination and more of these flowered in their first year

(Schaffner 2000 and personal communication). Only tetraploids have invaded North America, but these do not differ from European tetraploids in their colonization characteristics.

Being a larger plant is not necessarily the same as being a better competitor, and other factors can be involved in determining the competitive interactions among plant species. Turkington and Mehrhoff (1990) suggested that some plants compete better with long-established neighbors than with newer ones, and that plant communities may be more 'tightly knit' than previously envisioned. Callaway and Aschehoug (2000) found that diffuse knapweed, *Centaurea diffusa*, interacts differently with its plant neighbors depending on whether the neighbors are from North America (recently colonized) or Eurasia (the source of *Centaurea*). Root exudates from *Centaurea* had a greater impact on the newer neighboring grass species compared with Eurasian grasses. Thus, neighboring species with which knapweed had evolved were more competitive. Also, some introduced species may be able to use competitive mechanisms that are not present in the newly colonized communities to disrupt previously established interactions (Callaway and Aschehoug 2000).

It is clear that the question of whether introduced plants will be larger and more competitive in their new habitats will not be answered by broad surveys. The responses of plants to new environments will undoubtedly be influenced by their growth forms, their genetic backgrounds, their histories, their phylogenies and their neighbors. A variety of factors such as soil microbes and nutrient and mineral conditions can cause plants to differ in their growth and vigor. It is certainly true that some introduced plants reach much higher densities in exotic habitats than in their native habitats and many factors contribute to this success.

Are more diverse communities less vulnerable to invasion?

The notion that communities with increased species diversity are less likely to be invaded, has several origins. The first is the classic work by Elton (1958), who argued that simple communities are less stable and therefore more vulnerable to invasion. However, this prediction has not been supported. A review by Levine and D'Antonio (1999) showed that most studies contradicted the suggestion of Elton, that more diverse communities are more resistant to invasion, and instead showed that more diverse communities are more invaded (Figure 3.12). In these studies, it is usually the community following successful invasion that is considered,

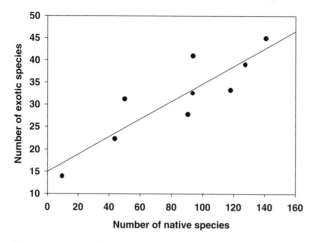

Figure 3.12. Number of native species and number of exotic species for shrubland sites. After Levine and D'Antonio's (1999) reanalysis of data presented in Fox and Fox (1986).

not that at the time of the initial species invasion. More recently, a positive correlation between the diversity of native and exotic species was observed by Smith and Knapp (2000) for 15 years of data from experimentally burned grasslands. This pattern also held for plant communities in Africa, with corrections for area effects (Stadler *et al.* 2000). In northwestern Kenya, zones with high richness of native species also had high richness of introduced species, mainly from Europe and the Americas.

Higher species diversity of communities might also indicate their degree of saturation. Under the equilibrium view, it is assumed that most plant communities are saturated, and therefore colonizing species would have to displace existing species. If more diverse communities are more completely saturated, they should be less easily invaded. The relationship between regional and local species saturation (Figure 3.13) can be used to determine if communites are saturated. Analyses by Caley and Schluter (1997) and Srivastava (1999) have shown that most communities are unsaturated. These results also disagree with the idea of diverse communites being less invasible.

Finally, the relationships among diversity, stability and community resistance attracted the attention of theoreticians, particularly in the 1960s and 1970s (review in Levine and D'Antonio 1999). The general basis of this theoretical work was that as more species used resources, the more

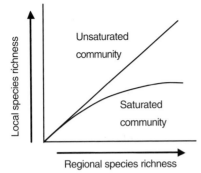

Figure 3.13. If communities are saturated the species richness of local communities should reach an asymptote above which no more species are added even though more species are available regionally. If communities are unsaturated the species richness of local communities should continue to increase with increasing regional species richness. After Krebs (2001).

efficient each species would have to be (see Box 3.1 for a further discussion of hypotheses on primary productivity, species richness and invasibility of plant communities). In diverse communities with highly efficient plant species it would be increasingly difficult for a new species to invade. MacArthur suggested that less diverse systems would have fewer linkages and this would make them more susceptible to fluctuations if one linkage was disrupted. These conclusions have been criticized, however, by those who propose that indirect interactions among competitors might make more diverse plant communities more invasible.

Arguments from theoretical studies can be made either way, on whether diversity will promote invasibility or resistance of plant communities. Mathematical models frequently indicate that diversity decreases invasibility (e.g. Case 1990). Whether this reflects nature or is a consequence of the model design can still be questioned. Models tend to be based on Lotka–Voltera equations and they consider communities in equilibrium. The composition of natural communities is much more likely to be dynamic and not in equilibrium.

Studies of experimental communities in which diversity is established as part of the manipulations are able to isolate the influence of diversity on invasibility. A study of grasslands in Minnesota, USA, found that invasibility was a negative function of diversity (Knops *et al.* 1999). In contrast Palmer and Maurer (1997) planted plots with one or five crop plants and found that more diverse plots were more frequently invaded. Plant communities dominated by the tussock forming sedge, *Carex nudata*, in

Box 3.1 · *Species richness, primary production, and plant invasion.*

Observations of species-rich communities show more introduced species (Tilman *et al.* 1996, 1997, Hector *et al.* 1999, Levine and D'Antonio 1999, Lehman and Tilman 2000, Stadler *et al.* 2000, Wardle 2001). However, experimental tests find that diversity reduces invasibility (Levine 2000 and review in Wardle 2001). Ecological interactions or sampling artifacts could account for the observed negative relationship between species diversity and invasibility of plant communities (Figure 3.14). The species-complementarity hypothesis predicts that more diverse ecosystems will contain species with a broader range of traits that will efficiently use resources. Productivity of these communities will be high and excess resources for introduced species limited. The 'sampling effect' hypothesis is based on the premise that experimental communities with more species will have a greater probability of containing highly efficient species. High species diversity will be correlated with high productivity. This species complementarity may be a consequence of the evolutionary history of co-occurring species (Sala 2001).

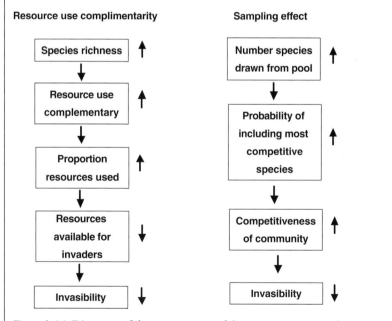

Figure 3.14. Diagrams of the components of the resource use complementarity and sampling effects hypotheses. From Wardle (2001) after Fukami *et al.* (2001).

Loreau and Hector (2001) compared the yield of plants in mono-cultures to that of plants in mixtures and concluded that species-complementarity can account for increased productivity with diversity. In contrast Wardle (2001) considered experimental studies and concluded that the apparent influence of plant species diversity on invasibility was due to 'sampling effect'. He proposes that the greater

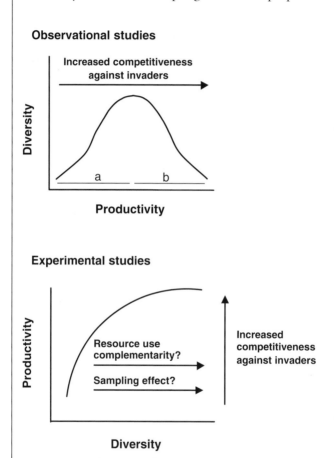

Figure 3.15. In observational experiments invasibility should be positively correlated with diversity over the productivity range indicated by 'a' because conditions will be less adverse for invaders. As productivity continues to increase 'b' greater competitive interactions would reduce the diversity of both resident and invading species and therefore again invasibility would be positively correlated with diversity. In experimental studies increased diversity leads to increased use of resources and reduced invasibility. However, results could be explained by either hypothesis. See also Figure 3.18 in regard to changes in resources and the invasibility of communities. After Wardle (2001).

invasibility with higher plant diversity in observational studies is associated with higher productivity of diverse communities. The negative relationship between diversity and invasibility of experimental studies is thought to be due to the inclusion of competitive dominants in more diverse communities. Distinguishing between resource use and sampling effects will require careful experimentation because they have similar predictions, as diagramed in Figure 3.15.

Kennedy *et al.* (2002) studied the effects of neighborhood properties of 147 experimental grassland plots in Minnesota on the invasion of 13 species of invasive plants by comparing plots with invading plants to randomly placed 'null' positions. Local species richness, neighborhood species richness, and crowding were related to decreased success of invading species. They conclude that at a local scale diversity in itself reduces the invasion of plants.

Few studies explicitly relate introduced species to productivity. Smith and Knapp (1999) showed that the richness of introduced species declined as primary production increased. In South America, where palatable African grasses have replaced previously forested sites, primary production and carbon sequestration are lower in the grassland than in native forests (Mack *et al.* 2000). Clearly the association between the productivity and resistance of communities to invasive species requires further study. Experimental manipulation of productivity through fertilization (Green and Galatowitsch 2002) or the addition of nitrogen fixing legumes could expose the relationships between the productivity, diversity and resistance of plant communities to invasion.

the South Fork Eel River in California showed a positive relationship between the species richness of tussocks and the incidence of invaders (Figure 3.16A) (Levine 2000). However, when the diversity of the tussocks was manipulated experimentally, the success of invaders introduced as small transplants was negatively related to the species richness of the tussock (Figure 3.16B). When equal numbers of seeds of invading grasses were added to sedge tussocks there was no relationship between community diversity and the establishment of introduced grasses. These conflicting results show how important the history of the community can be.

Lavorel *et al.* (1999) established artificial plant communities in Montpellier, France in which the number of species and the number of functional groups within three plant categories: annual grasses, annual

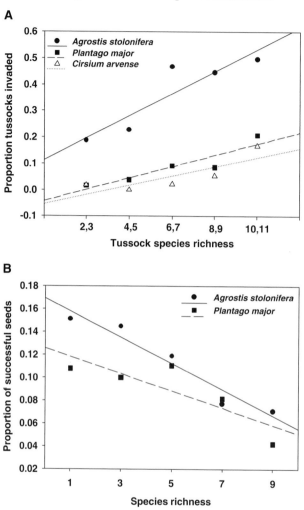

Figure 3.16. (A) Natural pattern of invasion relative to species richness on sedge tussocks by *Agrostis stolonifera* ($R^2 = 0.82$, $p = 0.04$), *Plantago major* ($R^2 = 0.89$, $p = 0.02$), and *Cirsium arvense* ($R^2 = 0.67$, $p = 0.09$). (B) Proportion of successful seeds of *A. stolonifera* and *P. major* on plots in which the species richness of native species was manipulated experimentally. After Levine (2000).

legumes and annual Asteraceae, were manipulated. They found no strong relationship with either the diversity of functional groups or species richness and invasibility from the natural seed bank. In experiments with potted plants in Australia, increasing density of co-germinating individuals reduced the establishment of *Echium*, only when a species that formed flat rosettes similar to *Echium* was included. They conclude that diversity

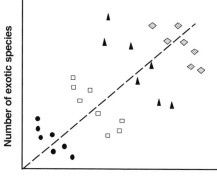

Figure 3.17. Hypothetical example of how invasion success and species richness of the native community within local areas as might be observed in small-scale experiments or observations (different clusters of points), and across a larger geographic scale covering a broader range of extrinsic factors. Within patches the relationship might be negative while on a broader spatial scale the relationship could be positive. Adapted from Shea and Chesson (2002).

needs to be measured in terms of functional groups, growth patterns, and in qualitative terms as well as quantitative terms, but rejected the prediction that increased diversity increased competition.

Covarying factors such as nutrients, seed numbers, and disturbance can influence the invasibility of communities and the success of invaders. In addition, the species diversity of a community could also influence the success of invading species if they include invasion promoters such as pollinators, soil fungi, nitrogen fixing bacteria, or organisms that disperse seeds. Observed relationships between diversity and invasibility are likely to indicate sites with good conditions for both native and introduced species. In contrast, experimentally created diverse patches are likely to reduce the success of introduced plants if cover is greater in more diverse plots. The relationship between the number of native species and the number of introduced species might be negatively related on a local scale but still appear to be positively related on a larger geographical scale (Shea and Chesson 2002) (Figure 3.17). The impact of propagule number will also vary with the distribution and availability of sources. With all of this potential for spatial and temporal variability it is not surprising that clear and consistent patterns do not emerge.

Invasions and fluctuating resource availability

If more productive habitats have higher species richness (Srivastava 1999) the role of resources on invasibility becomes of interest. Davis *et al.* (2000)

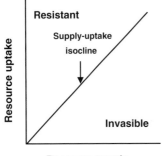

Figure 3.18. A diagram of the theory of fluctuating resource availability, after Davis *et al.* (2000). The invasibility of a community is related to the resources available and the level of uptake of these resources. Below the 'resistant' line communities are more invasible.

proposed that fluctuating resource availability is the key to understanding invasibility. The basis for this suggestion is that a plant community with unused resources such as water, nutrients, space, or light, will become more susceptible to invasion. The susceptibility of a community to invasion will be influenced by the supply of resources and the uptake of those resources (Figure 3.18). Any factor that makes resources more available, e.g. a wet spell, the input of nutrients or the removal of a species, will promote invasion by new species. This resembles Grime's model in which competition is less important following disturbance. But the concept is extended to include any increase in resources such as might occur. A corollary of this is that the invasibility of a community is not fixed, but will change with conditions. Davis *et al.* (2000) summarized some examples in which changing resources have influenced invasion (Table 3.2).

The proposed relationship between resource abundance and invasion generates several testable predictions. First, environments most likely to be invaded are those with periodic enrichment or an abrupt increase in the supply of resources such as after fire or flooding. Secondly, environments in which a decline in the uptake of resources has occurred will be susceptible to invasion. Thirdly, because conditions can change, no consistent relationship between the species diversity of a plant community and its susceptibility to invasion is expected. Finally, there will be no general relationship between the average productivity of a plant community and its susceptibility to invasion. What matters is the variation in productivity.

Table 3.2. *Studies showing a relationship of resource variation to invasibility of plant communities reviewed in Davis* et al. *(2000)*

Resource	Change	Reference
Nitrogen addition	Increased grass invasion	Huenneke *et al.* (1990)
Low nitrogen	Low invasibility of serpentine grasslands	Harrison (1999)
Nitrogen fixing shrub	Increased invasion of coastal prairie	Maron and Connors (1996)
Nutrient enrichment	Increased invasion of limestone grassland	Burke and Grime (1996)
Eutrophication	Increased invasion of Australian wheatbelt	Hobbs and Atkins (1988)
Increased water	Increased invasion of dry sites	Numerous studies cited in Davis *et al.* (2000)

The concept of 'niche opportunity'

Niche opportunity is defined by Shea and Chesson (2002) as 'the potential provided by a given community for alien organisms to have a positive rate of increase from low density'. This concept integrates the level of resource availability (as in the theory of fluctuating resources described above), escape from natural enemies, and the characteristics of communities into which invasion might occur into an integrated framework (Figure 3.19). Niches are defined as the relationships between organisms in the physical and biological environment in the context of space and time. The potential for a species to invade or its 'niche opportunity' will be determined by how it responds to changes in resources in a different manner to resident species or by using resources at times when the demand by resident species is low. Invading species are expected to have reduced impact from specialist predators and this could influence their niche opportunities. An aspect of community theory related to the potential for invasion is that, in 'mature' communities, species will be well adapted to each other and more resistant to invasion.

Shea and Chesson (2002) suggest four areas for future research: (1) studies of interactions between natural enemies and competitors in both native and introduced communities, (2) studies of invasion resistance considering time since community establishment, species pool, established species and functional diversity of species in the community, (3) development of new methods for data analysis in studies of covarying extrinsic factors, and (4) studies of changes in dominance patterns in

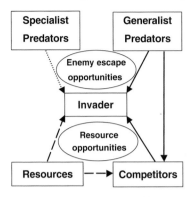

Figure 3.19. Components of the plant community that can influence the niche opportunity of an invasive species. Solid arrows indicate potentially negative impacts and dashed arrows potentially positive impacts. The dotted arrow indicates that the impact of specialist predators might be reduced on the invader as compared to other competitors in the community. Impacts can be directly on the invader or indirectly on competitors in the community. After Shea and Chesson (2002).

time and space with varying environments. These will require extensive and perhaps complex experiments. Shea and Chesson (2002), like Davis *et al.* (2000), call for considering invasion ecology to be a component of general population and community ecology. They are optimistic that invasion ecology will make contributions to community ecology, and we agree.

Ecological niche modeling

It is a complex matter to predict how emerging invaders will alter plant communities (Mack 1996). It is, however, agreed by all that a general, predictive theory of species invasions is still lacking (Peterson and Vieglas 2001). A recently suggested approach is 'ecological niche modeling' which attempts to identify those portions of a landscape that are habitable for invading species. Ecological niche modeling attempts to use information on species distributions to predict where species can maintain populations. Tests have been done, but only with insects and birds. This approach relies heavily on the ecologist's ability to model the niche. While niche modeling has some promise, its usefulness, or even the feasibility of applying it to plants, has not yet been explored.

A similar approach to niche modeling is 'climate matching', to predict where plants might invade. One model is CLIMEX (Kriticos and Randall

2001) that predicts the potential geographical range of an introduced species, and may be helpful in determining if a species will become invasive.

In conclusion, observed patterns of species invasion suggest a positive association between species diversity and invasibility of communities. Communities appear to be unsaturated and this does not agree with theoretical predictions. Some species of introduced plants may be more competitive in the exotic environment but this is not always the case. Communities are dynamic and changing use of resources may cause communities to vary in their susceptibility to invaders. In the next section of this chapter we consider how invasive species influence communities.

Part 2 – The effects of invasive species on plant communities and ecosystems

Plant communities change over time, and along successional gradients. Both native and introduced colonizing species arrive and may become established. The stability of communities can be measured over time, and through their ability to resist or respond to change by returning to their original state. The ability of a system to remain unchanged following a disturbance is known as **resistance**, while the ability of a system to return to its previous state following a disturbance is known as **resilience**. Stability, resistance and resilience are all 'emergent' properties of ecosystems and communities and are scale-dependent. McCann's (2000) review of the stability–diversity debate defines terms used in this branch of ecology (Box 3.2). He points out that while the theory here rests on the underlying assumption that populations are at equilibrium, there is little justification for the assumption (McCann 2000).

Feedbacks are fundamental to the regulation of change (DeAngelis 1992, Vitousek 1986). Negative feedbacks slow down a rate or process, and are ultimately stabilizing. Positive feedbacks accelerate rates or processes, and are ultimately destabilizing. Once a feedback has been created, it may be self-sustaining, and the removal of the factor that initiated it will not necessarily restore the system to its original state. Herbivores such as lesser snow geese (Bazely and Jefferies 1997) and moose (Pastor and Naiman 1992) may alter ecosystem processes via the creation of new feedbacks.

A basic question here is how the arrival of an invasive introduced species changes the plant community. This could be both by changing plant diversity or by modifying how the community functions.

Box 3.2 · *Definitions of stability after McCann (2000).*

Dynamic stability

Equilibrium stability	System returns to its equilibrium after a small perturbation. Does not vary without perturbation. Discrete measure.
General stability	Increases as the lower limit of population density moves away from 0. Under non-equilibrium dynamics a decrease in population variance is implied.
Variability	The variance in population densities over time, usually measured as the coefficient of variation ($CV = 100$ variance/mean).

Resilience and resistance stability

Equilibrium stability	System more stable when it returns to equilibrium more rapidly following perturbation.
General resilience	Rapid return to equilibrium/non-equilibrium solution after perturbation.
Resistance	Degree to which a variable changes after perturbation. Ability to resist invasion.

Effects of invasive plants on plant diversity

A simple expected outcome of the successful establishment of an introduced species, would be a negative correlation between plant density or cover and species richness. The negative correlation between the richness of true woodland species and increasing cover of nettles, *Urtica dioica*, in Belgian woodlots (Figure 3.20) illustrates this prediction nicely. However, this is a correlational relationship (Watt 1997). Such studies frequently do not take into account the disturbance history of sites, as this is often not known. Perhaps the native species were affected by disturbance, and in decline prior to the arrival of the species that later became 'invasive'. In this section, we look at the impacts of introduced species at the community and ecosystem levels in a search for ways of detecting causal rather than correlational relationships.

Some dramatic examples illustrate the potential for invading plant species to form vast monospecific stands. These species behave as 'blanket invaders' – carpeting the landscape and obliterating the vegetation that previously occupied these sites (Mack *et al.* 2000). The kudzu invasion in the southeastern United States is an example of this, as described in Chapter 2. It is almost unimaginable that such invaders have small effects,

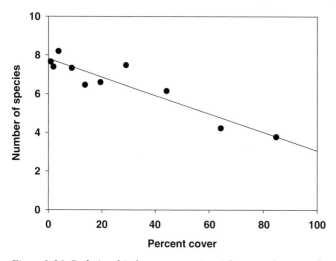

Figure 3.20. Relationship between species richness and cover of nettle, *Urtica dioica*. After Hermy (1994).

but their impacts are difficult to evaluate because data are rarely available both before and after invasion.

In temperate North America, purple loosestrife is 'invading' wetlands and spreading across the landscape. However, Anderson (1995) found little evidence that loosestrife outcompetes cattails and other wetland species. Blossey *et al.* (2001) report that loosestrife clearly forms monospecific patches and suggests that ecosystem-level effects such as changing water flow or wildlife habitat may be more important than impacts on species richness.

Another abundant invasive species in the forests of eastern North America is Amur honeysuckle, *Lonicera maackii*. Collier *et al.* (2002) evaluated the impact of this plant in secondary forests in southwestern Ohio by comparing species richness in sites invaded by honeysuckle over 16 years previously to recently invaded sites. Mean species richness for all species was 53% lower in plots below the honeysuckle crowns and cover was 63% lower there, and short- and long-term plots did not vary. Native shrubs are generally lacking from these forests and therefore honeysuckle is filling a new niche.

The impact of Amur honeysuckle on native vegetation has also been measured experimentally. Gould and Gorchov (2000) established experimental plots to compare the influence of honeysuckle on the survival and fecundity of three annual species, *Galium aparine*, *Impatiens pallida*,

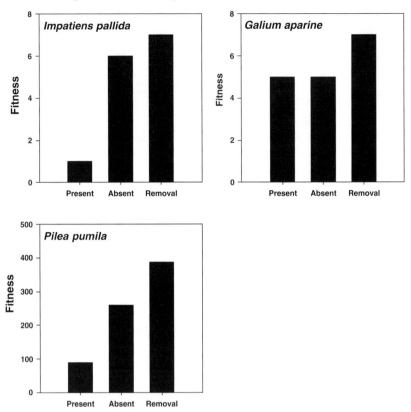

Figure 3.21. Fitness, the proportion of individuals surviving to reproductive age ×
fecundity for three annual species planted into plots with Amur honeysuckle,
without honeysuckle and from which honeysuckle was removed. Data from Gould
and Gorchov (2000).

and *Pilea pumila*. On one site they had three types of plots: present, absent
and removal. On another site they had only present and removal plots.
Seedlings of the annuals were planted in the plots and survival and fecundity were measured at the end of the summer. Removal of honeysuckle
increased the fitness of all three annuals (Figure 3.21). The response was
weaker in plots from which honeysuckle was originally absent and this
might indicate competition from other vegetation in these plots.

More of these experimental approaches to measuring the impact of
invasive species on the species diversity of native communities are required. Results are likely to vary among species and communities and
with experimental designs. The spatial scale will be important as well. In

such experiments, the number of species will be less likely to change than will species diversity that also considers the evenness of the distribution among species (see Appendix).

Effects of introduced species on ecosystem functioning

Experimental addition and removal of species to and from communities has demonstrated that some species, referred to as 'keystone species' (Paine 1966), can significantly alter ecosystems. However, most species additions show only weak impacts (Williamson and Fitter 1996a, McCann 2000). Similarly, most introduced species do not apparently have large influences on plant communities and ecosystems. However, some introduced species do reduce species richness directly or alter ecosystem processes to reduce species richness (Vitousek 1986). Many 'blanket invaders' may alter light levels through severe shading. For example, light levels in habitats colonized by the introduced prickly shrub, *Mimosa pigra*, were lower than in undisturbed native plant communities (Braithwaite *et al.* 1989). Thus, the introduced species that are of greatest concern are those that can change the structure of the ecosystem.

Vitousek (1986, 1990) identified three conditions which, if met by an introduced species, may alter the ecosystem: (1) invaders that acquire and use resources in a different way from native species, (2) invaders that alter the trophic structure of the invaded area, (3) invaders that alter the disturbance frequency or intensity. A number of species considered to be particularly invasive, including *Myrica faya*, saltcedar, *Tamarix* spp., and ice plant, *Mesembryanthemum crystallinum*, meet condition (1), in that they use resources differently from native species. *Myrica faya* was introduced to Hawaii in the late 1800s. This species is an early successional colonizer of young volcanic regions in Hawaii Volcanoes National Park. It increases overall soil nitrogen levels in what is a nitrogen-limited environment and this provides nitrogen for its own growth and that of other species. Prior to this there were no actinorrhizal, nitrogen fixing plants. There are clear indications that the altered nitrogen levels affect the native biota.

Saltcedar or tamarisk species, *Tamarix* spp., that were intentionally introduced to the USA from Eurasia (see Chapter 2), cover from 470 000 to 650 000 ha of native riparian forest habitat in 23 states in the USA (Zavaleta 2000b). *Tamarix* alters hydrological regimes by increasing evapotranspiration rates (see Zavaleta 2000a). It draws down the water table, dries desert springs, and lowers river flow rates and lake levels (Loope *et al.* 1988, Zavaleta 2000a,b). Tamarisk stands have been estimated to consume 3000 to 4600 m^2 ha^{-1} yr^{-1}, more water than the native vegetation that they replace. Zavaleta (2000b) points out that estimated levels of annual water losses due to tamarisk are similar to the total annual precipitation in the southwestern region of the USA, which is mostly less than 4500 m^2 ha^{-1} yr^{-1} and even less than 2000 m^2 ha^{-1} yr^{-1} in parts of this region.

In cases where species have radically different patterns of water uptake than native vegetation, flood regimes may also be altered, meeting condition (3) above by changing the intensity or frequency of disturbance. A wide range of non-indigenous woody species, such as *Lantana camara*, have been reported to change hydrologic patterns such as river flows (Le Maitre *et al.* 2002). Le Maitre *et al.* (2002) attributed the changes in flow rates of some South African rivers to the increased evapotranspiration rates of invading plants. In contrast, the reduced flow associated with rivers in the USA invaded by *Tamarix* has been attributed to its ability to trap sediment and cause extensive siltation of some river channels (Zavaleta 2000b).

The paperbark tree, *Melaleuca quinquenervia*, is considered to be a highly invasive species in South Florida, where it too was originally planted for erosion control (Turner *et al.* 1998). It is widely distributed and has become the dominant plant in many areas, where it replaces sawgrass, *Cladium jamaicense*, wetlands (Laroche and Ferriter 1992). It has also been labeled as a high user of water, and Westbrooks (1991) reported that it was planted as a 'swamp drying plant' (Westbrooks 1991). However, its impact on hydrological regimes through evapotranspiration rates that are reportedly 3 to 6 times that of native vegetation (Westbrooks 1991), has been questioned (Allen *et al.* 1997). The scanty data for *Melaleuca* come primarily from studies of potted plants and no field-based studies have been done to measure actual impacts. Extrapolating results from potted plants to field situations must be done with great caution (Allen *et al.* 1997).

The ability of some plants to secrete salts on their leaves can increase soil salinity levels (Vivrette and Muller 1977, Zavaleta 2000b). Both *Tamarix* and ice plant, *M. crystallinum*, draw up salts that are then secreted on their leaves (Vivrette and Muller 1977, Zavaleta 2000b). These salts subsequently wash off and increase surface soil salinity. Native riparian trees displaced by *Tamarix*, such as Goodding willow, *Salix gooddingii*, and Fremont cottonwood, *Populus fremontii*, are not salt tolerant and thus are not able to persist in these areas. Similarly, *M. crystallinum*, which has colonized many of the islands off the coast of California, has an impact on salt intolerant species. After the plants die, the rain washes salt crystals into the soil, and osmotic levels increase, creating saline conditions under which the seedlings of most plant species cannot survive (Vivrette and Muller 1977).

Examples of invaders altering trophic interactions are usually of animals (Vitousek 1990). However, Braithwaite *et al.* (1989) demonstrated that invasion by the prickly shrub, *Mimosa pigra*, was associated with declines in some of the native fauna of coastal sites in Northern Australia. They predicted that in the future, densities of the introduced water buffalo would decline as wetland areas dominated by *Mimosa* underwent reduced food availability. This kind of response to a plant invader that is heavily defended against herbivores would be predicted (Myers and Bazely 1991). This is one means by which plant invaders could potentially have a 'bottom-up' impact on upper level predators. Because water buffalo are introduced, their reduction would not be a bad thing.

In the case of *Tamarix*, the impacts on the trophic structure may be occurring through its negative impacts on native wildlife, by its drying up of springs and its lack of palatable fruit and seeds (Loope *et al.* 1988, Zavaleta 2000a,b). Populations of yellow warbler, *Dendroica petechia*, Bell's vireo, *Vireo bellii*, and several other bird species that depend on cottonwood, willow, mesquite and other native plants have been reduced or extirpated in *Tamarix* infested areas (DeLoach 1990). On the other hand, the biological control of *Tamarix* was delayed because a rare and endangered bird species, the southwestern willow flycatcher, *Empidonax trailii extimus*, that formerly nested in willows, adopted *Tamarix* as a nesting habitat (Zavalata 2000a).

The classic example of an invader that meets condition (3), the alteration of disturbance regimes, is that of the introduced grass *Bromus tectorum* in western North America. It has come to dominate an area of approximately 200 000 km^2 in the 'intermountain' area since 1863 (Novak and Mack 2001). This annual grass is reported to increase the probability, extent and severity of fires (D'Antonio and Vitousek 1992) (Figure 3.22). Other introduced grasses in North America, such as *B. madritensis* ssp. *rubens*, *Schismus arabicus* and *S. barbatus*, which are found in the Mojave Desert, have also been implicated in the alteration of fire regimes (Brooks 1999). Another grass, *Andropogon*, has been identified as a 'fire-enhancer' in Hawaii, where the fires increased 100-fold following the removal of feral goats (Vitousek 1986). The ability to 'enhance' fire regimes is not restricted to grasses, which increase the fuel load. While

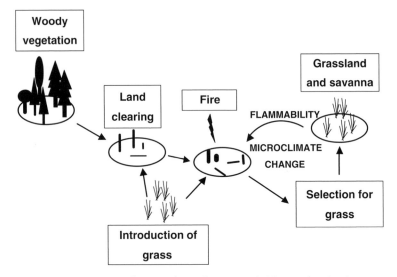

Figure 3.22. Invasions of grass to logged areas can initiate and maintain a grass–fire feedback system that prevents the regeneration of native woody species. Adapted from D'Antonio and Vitousek (1992).

Melaleuca quinquenervia's ability to moderate hydrologic regimes is not clearly established, it does have an impact on fire regimes (Turner *et al.* 1998). Crown fires in *Melaleuca* burn hotter than sawgrass fires, to the detriment of a number of native species (Turner *et al.* 1998).

The examples above all describe cases where the ecosystem moves to an alternative state following establishment of the non-indigenous species. It has yet to be explicitly demonstrated that invasive species can create self-sustaining positive feedbacks (DeAngelis 1992) that continue to operate, even when the organisms are removed from the habitat. Vitousek's (1986, 1990) research on the invasive plant *Myrica faya* clearly demonstrated that introduced plant species could accelerate soil nitrogen fixation and increase overall nitrogen levels (reviewed in Vitousek 1990). It remains to be resolved what would happen to these ecosystems if the *Myrica* were removed. Will the communities revert to native ones? Reever *et al.* (1999) in the same region found that following reductions in nitrogen enrichment, the plant community composition did not change, although overall biomass declined. It would seem likely that the effects of *M. faya* on the soil could be very long-lasting, even if the plants were eradicated. Similarly, Milchunas and Lauenroth (1995) found an 'inertia' in species composition of plots in a long-term experiment where water and nitrogen were applied to a prairie community. This inertia particularly applied to the plots that had been watered. Long after the treatment had stopped, the community, which had been slowly shifting back toward conditions in the pre-treatment control, reverted toward the treatment community. This suggested that 'biotic' effects, unrelated to the original treatment, were in place.

Ehrenfeld and Scott (2001) point out that more subtle or indirect effects of introduced species on ecosystems have received relatively little attention. Gordon (1998) reviewed 31 invasive species in Florida, and concluded that 12–20 species (39–64%) were potentially altering ecosystem properties such as geomorphology, hydrology, biogeochemistry, and disturbance.

Invasive species and the soil

Recent research has examined the ability of species other than the 'classic' invaders mentioned above, to alter soil conditions in ways that favor increased colonization. These alterations are the equivalent of the 'facilitation' model of succession described by Connell and Slatyer (1977) in which colonizing species alter habitat conditions to make them more favorable for colonization by other species. The process of soil formation that may occur during facilitation, is a natural step in primary succession. It has been observed to occur in Great Lakes sand dune communities, when red cedar, *Juniperus virginiana*, creates 'nucleation' sites where other colonizing species are able to become established (Yarranton and Morrison 1974). As Ehrenfeld and Scott (2001) point out, when invasive species alter soil processes, they may create changes that have the potential to radiate through the ecosystem.

Soil properties that change in the presence of introduced species, include soil biota – bacteria and fungi (Belnap and Phillips 2001), and nitrogen cycles (Mack *et al.* 2000, Ehrenfeld *et al.* 2001, Evans *et al.* 2001, Scott *et al.* 2001). The most important observation from these studies was that changes were species specific. In some cases, the nitrogen content of litter was greater than that of the displaced native vegetation (Ehrenfeld *et al.* 2001) and in other cases it was lower (Scott *et al.* 2001). However, there was enormous seasonal variation (Scott *et al.* 2001) and nitrogen cycling rates were also affected by disturbance regimes (Mack *et al.* 2000). This agrees with studies of the impacts of herbivory on nitrogen cycling; in some cases nitrogen availability is accelerated (Bazely and Jefferies 1997) and in other cases it is slowed down (Pastor and Naiman 1992).

Conclusions

In the first part of this chapter we show that plant communities and ecosystems are highly dynamic. They follow trajectories of change over time that are not fully predictable. Multiple biotic and abiotic factors determine the direction of change in plant communities, and there are many ways in which they can interact to produce different outcomes. This is why disturbances such as herbivory and fire can have positive or negative effects on introduced species. It is also why the effects of interspecific competition between introduced and native species are strong in some years, but absent in others, depending on weather conditions. Trying to predict whether an introduced plant species will become an invader remains an elusive quest. It depends on many modifying effects of the environment.

Predicting those communities that will be invaded is also difficult. Diversities of introduced and native species are frequently correlated. However, experimental tests of diversity and invasibility of plant communities show that the resulting relationships are influenced by the type of plants and the particular conditions of resource availability.

A main conclusion to be drawn from the second part of this chapter, is that the impact of an introduced species on plant communities and ecosystems can be enormous if it acts as a 'keystone species'. Then it can overcome the stabilizing effects of high species and functional group diversity and the buffering effects of food-web interactions. This is most likely to occur if the species alters ecosystem processes and creates self-sustaining feedbacks (Vitousek 1986, 1990). Areas for further research include determining whether the most notorious invaders all share this common characteristic as predicted by Gordon (1998). A final message is that species are not equal and their life history, morphological, and biological characteristics must be considered in experiments and in observational studies attempting to reveal general patterns of the impacts of introduced species on native communities. These factors are considered in the following chapters.

4 · Predicting invasiveness from life history characteristics

What are life history traits?

Traits that affect the fitness of individuals directly are 'life history traits' (Stearns 1992). For plants, these include size and number of seeds, seed dormancy, growth pattern and the size or age at first reproduction. Plant species differ in how and how often they reproduce, how rapidly they grow, and their physical structure. These characteristics can influence the evolution, adaptation, population growth and the population dynamics of species. Some characteristics associated with variation of life history patterns are given in Table 4.1. These same characteristics could contribute to whether an introduced plant species might become established in a new area and then whether it might invade and outcompete the native plants. We are interested here to know if invasive plants have life history characteristics that contribute to their success. We will first consider some of the variation in life history characteristics that might be important.

The growth rate and time to first reproduction of plant species determine how quickly a population can grow. Characteristics that influence plant survival will determine the persistence of a species in a habitat. The growth form and root structure of a species will influence its ability to compete. Characteristics that influence whether plants outbreed, inbreed or reproduce asexually will determine the levels of genetic variation that they maintain, and thus how rapidly they might respond to changing selection and varying environmental conditions. The persistence of seeds in the soil can buffer plants against poor conditions. Some species have large and persistent seed banks, while for others, most seeds germinate in the next growing season.

Vegetative reproduction

Not all invasive species reproduce only by sexual reproduction and seed dispersal alone. Some plants both flower and grow vegetatively by sending

Table 4.1. *Some of the characteristics of plants that vary among species and potentially influence their reproduction and population dynamics*

Characteristics	Types
Life stages	Seed, seedling, rosette, flowering
Reproduction	Clonal or sexual
Breeding	Outcrossing or selfing
Growth	Annual, short-lived perennial, perennial
Pollination	Self, wind, insect, other
Seeds	Many small, few large
Seed bank	Long-lived or short-lived
Growth form	Herbaceous, woody – tree or shrub
Flowering	Semelparous (once) or iteroparous (more than once)
Root system	Tap root or root mat

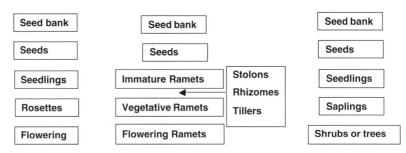

Figure 4.1. Three stylized life history patterns of plants.

out structures to initiate new individuals, and some plants lack sexual reproduction entirely. Three basic life history patterns are diagramed in Figure 4.1. For vegetatively growing plants the total plant is known as the genet and ramets are the individual portions of the plant that develop roots and are eventually capable of independent existence. Clonal growth is accomplished by different structures in different species. Grasses produce tillers from lateral meristems, other vegetatively reproducing plants produce stolons that are above ground, lateral stems, or rhizomes, that are below-ground lateral stems (Table 4.2). Plants capable of both sexual and vegetative growth seem to have many adaptive advantages over those with a single reproductive strategy; however, they do not dominate plant communities.

Table 4.2. *Types of asexual reproduction by plants*

Type	Characteristics
Agamospermy	Seeds without fertilization but pollination required
Vegetative	
Stolons and runners	Stems growing on soil surface with adventitious roots producing new shoots
Rhizomes	Stems growing underground with reduced leaves
Tubers	Thickened rhizomes mostly for storage and axillary buds
Bulbs	Large underground buds composed of stem and modified leaves
Corms	Underground stems covered with leaf bases
Roots and stems	Suckers or shoots from underground roots
Leaves and adventitious buds	Plantlets arising from leaf margins or tips
Fragmentation	Dispersal of vegetative plant parts

Adapted from Abrahamson (1980).

Time to first reproduction

The time to first flowering is an important determinant of the rate at which a population can increase (see Chapter 5). This can be both a general trait of the plant species – annual, biennial, short-lived perennial, long-lived perennial – and a characteristic of the individual plant in relation to other members of the population. Most plants are annuals or perennials and biennials make up only about 10% of species. However, these can be very common in some situations and many weeds are biennials or short-lived perennials (Silvertown 1983). The biennial life history of initially becoming established as a rosette and then flowering the next year may be a good strategy for utilizing sites which are suitable following disturbance, but become less so with succession (Meijden *et al.* 1992).

Growth versus flowering

For some species, the age at first reproduction of individual plants is very plastic and strongly influenced by environmental conditions. In this case, how rapidly the plant grows and develops from the seedling to the flowering stage determines the age at first reproduction for that individual. The rates of growth and development, in turn, are influenced by the environment; drought, crowding or poor soil quality can all prolong the

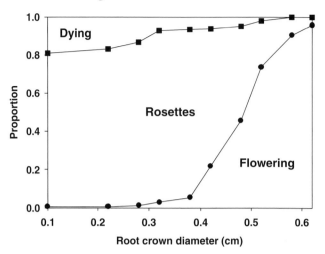

Figure 4.2. The fate of diffuse knapweed rosettes in the field in relation to their root crown diameters. The top line indicates the proportion of rosettes and the bottom line is the proportion of plants flowering. After Powell (1990b).

age to first reproduction. For other species, the time to first flowering is determined by the age of the plant, but when a particular individual flowers in the season may be modified by its genes as well as the environment. Plants in populations with genetic variation for the time of flowering can respond to selection. In increasing plant populations, such as those of newly established introduced species, early flowering may be selected for, since offspring will have a good opportunity to survive and reproduce. In stable populations, however, plants that are able to persist and grow will be more likely to be successful.

Short-lived perennial plants often show phenotypic plasticity in the time to flower and this is influenced by environmental conditions. For diffuse knapweed, *Centaurea diffusa*, an introduced rangeland weed in North America, whether a rosette matures into a flowering plant depends on its size as indicated by the root crown diameter (Figure 4.2). In a clever experiment Powell investigated the impact of crowding on the survival and flowering of rosettes (Figure 4.3). Plants in the center of the plots planted as a wagon wheels with eight spokes were crowded while those toward the edge were spaced at distances of 16 cm. Mortality was highest among the most crowded plants and the proportion of the plants flowering was the highest among the least crowded plants.

Whether plants should delay flowering or reproduce at a smaller size involves a tradeoff between growth and mortality. Prolonged growth

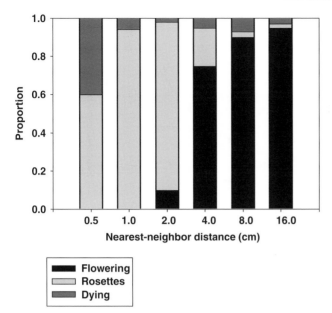

Figure 4.3. The fate of rosettes of diffuse knapweed subject to varying levels of crowding as indicated by the nearest neighbor distances. Dying (darkly shaded), persisting as rosettes (light shading) and flowering (black shading). After Powell (1990b).

exposes plants to higher risk of death as they are growing but the greater seed production of larger plants can be advantageous (Figure 4.4). Tansy ragwort, *Senecio jacobaea*, lives in open and disturbed sites in both Europe where it is native, and in the Neo-Europes – North America, Australia, New Zealand – where it has been introduced. In all of these areas, tansy ragwort can reproduce as a biennial or short-lived perennial. The population ecology of tansy ragwort has been studied in its native range in sand dunes on the coast of The Netherlands (Meijden and van der Waals-kooi 1979). After germination in the autumn or spring, the plant forms a rosette, and in the spring or early summer of the next year it can form a flowering stem. Once the plant has flowered it dies. This life history is known as monocarpic, reproducing once. The probability that a plant will flower in a particular summer depends on its size in the spring. In these populations mortality is high for small rosette plants, but if they flower they produce fewer flower heads. Survival is high for large rosettes and when they develop to flowering plants they produce large numbers of flowers. Tansy ragwort also exhibits clonal growth. Damage

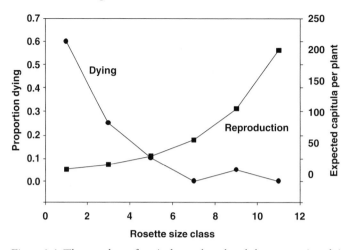

Figure 4.4. The number of capitula produced and the proportion dying in relation to the diameter of the rosette at the beginning of the flowering season for tansy ragwort growing in dunes in The Netherlands. Data from Meijden and Waals-kooi (1979).

by herbivores to tansy ragwort can delay flowering and stimulate clonal growth. This response was greater in larger rosettes (Figure 4.5). Thus tansy ragwort is well adapted to the changing and uncertain conditions it faces in the sand dune habitat. These reproductive characteristics have undoubtedly contributed to its success as an invader of open areas in other parts of the world.

Other selection pressures can influence the time of first reproduction of flowering plants. Among these are the activities of pollinators and herbivores. How responsive flowering time is to selection is determined by the degree to which the variation in flowering time of a plant species is genetically controlled, and how flexible the plants are to environmental conditions. Changes in flowering time in the presence of selection indicate genetic control. Two interesting field situations illustrate this relationship.

Impatiens pallida is an annual herb of deciduous forests of the eastern United States. Populations in woodlands in Illinois vary in flowering times between populations growing on the edge and in the interior (Schemske 1984). Herbivore pressure on plants in the interior during mid-summer apparently selects for earlier flowering. Along the edge of the forest, plants were large and lived at high densities. In the interior, plants were small and their densities low. While a variety of conditions differ between forest

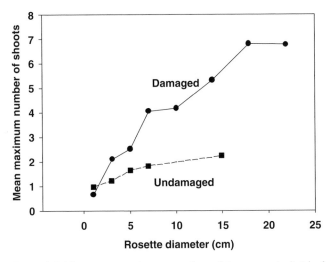

Figure 4.5. The mean maximum number of shoots per individual tansy ragwort for undamaged and damaged plants. Damaged plants were cut off at ground level. After Meijden and Waals-kooi (1979).

edges and the interior, the level of feeding by host specific leaf beetles, *Rhabdopterus praetexus*, was the most obvious difference. The beetles destroyed a high proportion of the forest dwelling plants. Plants grown from seeds of interior plants flowered significantly earlier than those from edge plants under the same controlled conditions. This adaptation allowed interior plants to produce seed before suffering the impacts of predation. Seed production was negatively correlated with flowering time and thus early flowering plants would only be at a selective advantage in the environment with early mortality.

Insects do not always select flowering time of plants in a directional manner, and a tradeoff between pollinator abundance and seed predators can maintain variation in flowering time in some species. For the shrub *Vaccinium hirtum*, a lack of pollinators in the early spring reduces seed set for early flowers (Mahoro 2002). Later in the spring pollinators are more common and seed production is higher for flowers blooming at that time. However, there is a catch. Later in the spring flowers are attacked by two insect species, a fly and a weevil, and this reduces their reproductive success (Figure 4.6).

Age at first reproduction is related to successful invasion by introduced plants. It was the only life history characteristic to differ significantly between native and invasive plant species in an analysis by Williamson and

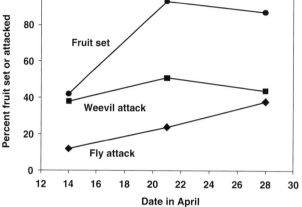

Figure 4.6. The impact of pollinators is indicated by the percentage of open flowers that produced seeds compared with experimentally pollinated flowers over two weeks in April. Lower values indicate insufficient pollinators. The percentage of flowers attacked by unidentified fly larvae and a weevil, *Mecysmoderes* sp., increase during April. Data from Mahoro (2002) are from 1999 at one site. Fruit set was higher in 2000.

Fitter (1996b) of the Ecological Flora Database in Britain. Of 26 quantitative and qualitative characters investigated, morphological characteristics such as life form, height, leaf area and plant spread (plant height greater than plant width) generally varied more than life history characteristics between newly established and native plants.

Seed germination and dispersal

As with time to first reproduction, factors associated with seed production and seedling establishment are characteristics of plant species as well as of individual plants. These characteristics are determined by both genetic and environmental influences. As will be illustrated below, small seeds and frequent large seed crops characterize invasive plants. In Chapter 5 variations in seed production in response to environmental conditions are described. In addition to the number of seeds produced, the survival of seeds and their successful germination are crucial to the ability of plants to invade new areas and to their population dynamics over time. Dispersal of seeds determines plant distribution and spread. The components of these interactions are outlined in Figure 4.7.

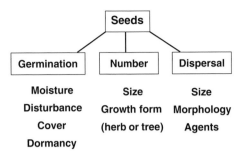

Figure 4.7. Factors influencing seed production and seedling establishment.

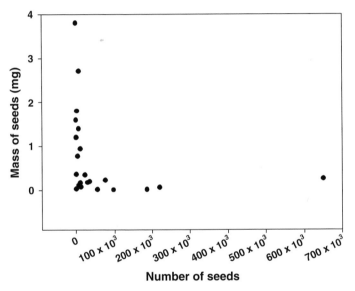

Figure 4.8. Relationships between seed size and the number of seeds produced by different plant species. Data from Salisbury (1942).

Plants are faced with tradeoffs in allocating resources among growth, survival and seed production. Plant species vary in the average size of seeds they produce (Figure 4.8) and a range of seed sizes also occurs among invasive species. For example, purple loosestrife produces millions of tiny (c. 0.5 mm diameter) seeds per plant, and the legume Scotch broom, *Cytisus scoparius*, produces thousands of larger seeds (c. 5 mm diameter). Crawley (1997a) describes some of the tradeoffs related to seed size. Larger seeds may be more attractive to predators on the negative side, and may be more competitive as seedlings on the positive side. Larger seeds may germinate more rapidly than small seeds. Thompson *et al.* (1993) showed for 53 British herbs that small, round seeds tended to remain in the soil longer than did larger, more elongate seeds. This pattern did not hold up for a sample of Australian species (Leishman and Westoby 1998) although comparisons here could be influenced by other differences between the two areas.

Seed banks

Seed dormancy provides a strategy for plants to cope with environmental variation. Seeds that delay germination accumulate in the soil or are eaten by predators. The accumulation of seeds in the soil forms the seed bank. The size of the seed bank will be determined by the proportion of seeds that fail to germinate when conditions first become appropriate, and the level of seed predation that occurs. Therefore, seed dormancy could result in a potentially large seed bank for a particular species, but if seed predators remove a high proportion of the seeds, the actual seed bank could be small. The existence of a seed bank spreads germination over good and bad years and is an excellent adaptation for plants living in unpredictable environments. Areas with hot dry conditions such as savannas and deserts have a high frequency of plant species with dormant seeds while moist areas, such as rainforests, have fewer species with dormant seeds (Baskin and Baskin 1998). On the other hand, annual grasses in Mediterranean climates have only a short innate dormancy, long enough to get them through to the next period of reliable rains and favorable growth temperatures (Groves 1986a).

Germination of purple loosestrife seeds, as in many other plants, is determined by temperature and exposure of the seeds. Seeds buried 2 cm deep do not germinate, but can remain viable for several years and therefore can respond to any disturbance that brings them to the surface (Welling and Becker 1990). Competition with a second introduced species, reed canary grass, *Phalaris arundinacea*, influences the germination of purple loosestrife seeds in North American wetlands. While 53% of sown loosestrife seeds became established as seedlings in plots from which grass was removed, no seeds established in plots with reed canary grass (Rachich and Reader 1999). Seed banks under Scotch broom plants vary from 400 to 27 000 seeds m^{-2} in exotic habitats and 460 to 10 000 seeds m^{-2} in native habitats (Rees and Paynter 1997). Again, soil disturbance stimulates germination of seeds in the seed bank.

Maintaining a seed bank might characterize invasive species, but this is not always the case. Two plant species of European origin that are invasive weeds in North America and are well studied in Europe are *Cirsium vulgare* and *Cynoglossum officinale*. These biennial species were studied in sand-dune areas of The Netherlands (de Jong and Klinkhamer 1988). For both species, persistent seed banks were small; 2% of the seeds of *Cirsium* failed to germinate but were still viable after the first spring and for *Cynoglossum* approximately 5–10% delayed germination. For both species, seed germination was strongly influenced by disturbance in experimental plots to which seeds were added, but not in plots without additional seeds (Figure 4.9). Therefore, in their native habitats seed banks were not an important aspect of the population dynamics of these species. However, both of these plant species used delayed flowering as a mechanism for persisting through bad years. For *Cynoglossum* the proportion of rosettes one year and older that delayed flowering was 75–95% and for *Cirsium* delayed flowering occurred for 53–80% of rosettes.

A second example of seed dormancy involves Scotch broom, an invasive shrub in the Pacific Northwest of North America. Parker (2000) estimated seed dormancy of broom in several locations by putting seeds out in mesh envelopes and collecting them at different time intervals. In prairie populations dormancy was approximately 40–45% and in urban populations it was 72–74%. Seeds are known to persist for a long time in laboratory conditions, but in the field predation and removal by ants have major impacts on the potential seed bank. Germination rates, and seedlings per seed, for broom differed from the patterns for *Cynoglossum* and *Cirsium* described above. Germination rates were quite low, generally less than 1%, and differed among populations and in relation to the stage of invasion of broom (Parker 2000, Figure 4.10). In keeping with the pattern of dormancy, germination was lower in the urban populations than the prairie populations. Whether this indicates a genetic difference between populations or different environmental conditions is not known. The germination rate differed with the history of the populations and was lowest in the center of the populations, decreasing toward the edge of invasion. This may be related to variation in the density of the populations.

Disturbance and seed persistence

If adult longevity and seed dormancy are alternate mechanisms for coping with an uncertain environment, short-lived plants are expected to have

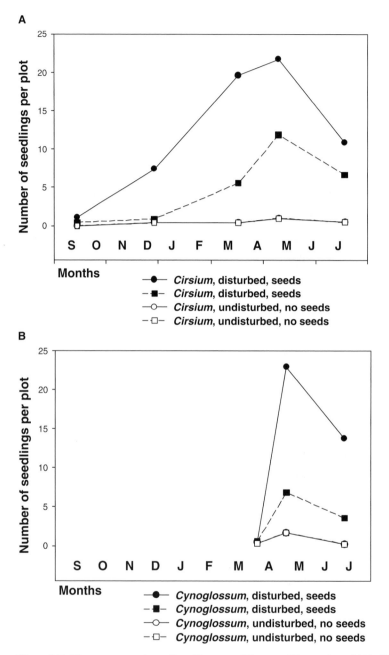

Figure 4.9. The mean number of seedlings per 30 cm × 30 cm plot of (A) *Cirsium vulgare*, (B) *Cynoglossum officinale* in disturbed (circles) and undisturbed (squares) plots with (solid symbols) or without (open symbols) seeds sown, in the sand dunes at Meijendel. After de Jong and Klinkhamer (1988).

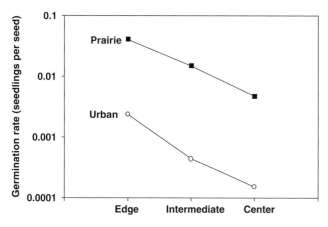

Figure 4.10. Mean germination rates (seedlings per seed) of Scotch broom in populations in urban (squares) and prairie (circles) sites in Washington. Three stages of invasion are evaluated by the center, intermediate and edge locations within the populations. After Parker (2000).

long-lived seeds. After correcting for seed size Rees (1993) found that this pattern holds for British plants. Similarly, Thompson *et al.* (1998) report that for plants in northwest Europe seeds of annuals and biennials tend to be more persistent than those of related perennials. However, no pattern between these traits and seed size was apparent. Rees (1993) compared seed size and dispersal to seed persistence and found that large-seeded plants and species with efficient dispersal of seeds in space have reduced seed dormancy. Large seeds are expected to produce more vigorous seedlings that are better than small seeds at becoming established in unfavorable environments. Dispersal spreads the risk of offspring among environments that may vary in quality. Both having a good potential to establish (large seeds), and spreading the risk through seed dispersal, should reduce the selection on seeds to remain dormant. Mechanisms of dispersal for seeds of different sizes may vary. Small seeds may be wind-dispersed and therefore they are predicted to have longer dispersal distances and reduced seed dormancy (Rees 1997). Dispersal of small seeds may also be passive and relatively short-range, and these

Table 4.3. *Characteristics of four 'biennial' plants and their establishment shown by the number m^{-2} of rosettes in bare and vegetated plots in a one-year-old field*

Species	Seed weight (mg)	Seeds/plant	Dispersal	Rosettes Bare	Rosettes Vegetated
Verbascum thapsus	0.064	100 000	Passive – 11 m	10	0
Oenothera biennis	0.20	6000	Passive – 4 m	21	0
Daucus carota	1.04	1800	Wind – 15 m	112	20
Tragopogon dubius	6.84	90	Wind >250 m	39	10

Data from Gross and Werner (1982).

would be predicted to have increased seed dormancy. On the other hand, large seeds will be efficiently dispersed by animals, and therefore these too would be expected to have little dormancy. In summary, although there are some patterns to seed morphology, dispersal, and germination, many exceptions exist.

Thompson *et al.* (1998) also used data from the northwest European flora to test the hypothesis that plant species living in disturbed habitats, mine tailings, and arable land have long-lived seeds compared with those in more stable sites, woods and pastures. They found that increased disturbance is associated with increased seed persistence, but that persistence was not always associated with reduced seed size. They attributed the lack of association of seed size and persistence to other physiological and habitat characteristics that might influence the probability that a seed will be buried, such as plowing arable land.

The relationship between seed longevity and rates of local extinction was also tested using the northwest European flora (Stöcklin and Fischer 1999). They compared the persistence of plants between 1950 and 1985 in 26 intact remnants of calcareous grassland. Rates of extinction were lower for plant species with seed longevity >5 years (33% extinct) compared to those with seed longevities of 1–5 years (61% extinct) and <1 year (59% extinct).

In a field study of these issues Gross and Werner (1982) investigated the colonizing abilities of four species of 'biennial' plants common to abandoned agricultural fields in northeastern North America. When seeds were added experimentally, all four species could colonize bare soil, but only large-seeded species became established in vegetated plots (Table 4.3). The seedlings of the large-seeded plants had more vertical leaves than did the small-seeded plants. This could facilitate their establishment in vegetated areas. Gross and Werner observed that while the large-seeded species became established in both one-year and 15-year-old fields, these species were generally not found naturally in newly abandoned fields. They suggest that this is because the seeds of these plants do not live more than one or two years. The two species with large seeds had longer dispersal distances and were better able to become established. Predation of the seeds of these two species differed, however. To quantify predation on seeds of *Daucus carota* and *Tragopogon dubius* Gross and Werner set out seeds in Petri dishes. Within 24 hours 73% of the larger *Tragopogon* seeds were removed while only 10% of the smaller *Daucus* seeds were removed over the same time. These experiments demonstrate some of the tradeoffs that influence the ability of plants to colonize new sites.

Seed size and seed predation

Another experimental study demonstrates the relationships between seed size, seed predation, and ground cover on germination of native and introduced plants in an old pasture in Ontario (Reader 1993). Seeds ranging in mass from 0.06 to 12.2 mg were planted in field plots with or without vegetation removal, and with and without predator exclusion cages. Excluding seed predators did not affect seedling emergence for the plant species with seeds <0.14 mg but did for those with larger seeds (Figure 4.11). Removing the vegetation significantly increased seedling emergence for the four plant species with the smallest seeds and for two species with larger, but hard coated seeds. Curiously, six plant species with larger seeds only responded to vegetation removal in the presence of predators. Under natural conditions ground cover and seed predation may provide conflicting pressures on small and large seeds and result in equal seedling emergence for small- and large-seeded plant species.

Thus the relationships between seed and plant population characteristics can be variable and comparative studies differ depending on the types of plant species considered, and the phylogenetic relationships among the plants studied. These are summarized in Table 4.4.

Seed dynamics can influence the maintenance and growth of plant populations, and the impacts of these can be elucidated by comparing closely related plant species living in the same environment. Exotic, perennial grasses have been used in North America to prevent erosion, reduce the spread of alien annual grasses and to rejuvenate overgrazed areas by providing grazing tolerant species. One of these introduced species is *Agropyron desertorum*. The demographic characteristics of this species were

Table 4.4. *Summary of relationships between plant and habitat characteristics and the persistence of seeds*

Characteristic	Response	Reference
Short plant life	Long-lived seeds	1,2
Large seeds	More predation	1,3,4
	Better establishment	1,3
	Short seed persistence	1,5
	Dispersal variable	1
Efficient dispersal	Wind – small, short-lived seeds	6
	Animals – large, short-lived seeds	6
Disturbed habitats	Long-lived seeds	2
	Increased establishment of small seeds	3
Seed persistence	No consistent pattern with size	2
Hot dry habitat	Long-lived seeds	7
Moist habitat	Short-lived seeds	7
Mediterranean habitat	Grasses with short-lived seeds	8
Reduced extinction rate	Long-lived seeds	9

References: 1. Rees (1993), 2. Thompson *et al.* (1998), 3. Reader (1993), 4. Gross and Werner (1982), 5. Thompson *et al.* (1993), 6. Rees (1997), 7. Baskin and Baskin (1998), 8. Groves (1986a), 9. Stöcklin and Fischer (1999).

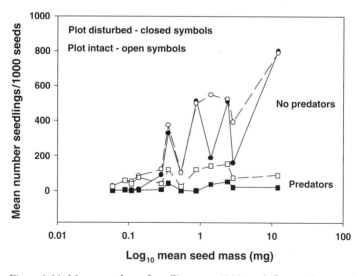

Figure 4.11. Mean number of seedlings per 1000 seeds for 12 plant species (*Solidago canadensis, Hieracium pratense, Hypericum perforatum, Poa pratensis, Chrysanthemum leucanthemum, Mentha arvensis, Taraxacum officinale, Daucus carota, Medicago lupulina, Centaurea nigra, Echium vulgare, Tragopogon dubius*) in experimental plots with and without vegetation and seed predators. Data from Reader (1993).

Table 4.5. *Comparison of the seed and seedling characteristics of two tussock grass species in Utah*

Characteristic	Agropyron desertorum (exotic)	Agropyron spicatum (native)
Seeds – wet year	1772 m^{-2}	26 m^{-2}
Seeds – dry year	1037 m^{-2}	0
Dispersal	Spring to next spring	Early to mid-June
Seed bank	Approximately $300–600 \text{ m}^{-2}$	Few seeds found
Seed predation	Half removed in 24 h	Half removed in 24 h
Seedling emergence	7.2% of 1425 seeds/plot	2.8% of 1425 seeds/plot
Summer seedling mortality	85%	86%

After Pyke (1990).

compared with the native *A. spicatum*, by Pyke (1990). Following herbicide spraying 12 years previously, the site in northwestern Utah was seeded with *A. desertorum* and by 1985, 38% of the canopy cover consisted of this introduced species and 12% of the native *A. spicatum*. Characteristics of the two species are given in Table 4.5. Greater seed production, a larger seed bank and greater retention of seeds in inflorescences by the introduced species allowed carry over of seeds from one year to the next and buffered the population from severe drought conditions that occurred in one year of this study.

The demographic factors associated with seeds favor *A. desertorum* and explain its maintenance and spread into native grasslands formerly dominated by *A. spicatum*. It is interesting that *A. desertorum* was initially introduced into the Intermountain West of North America in 1932 and was widely sown into this area until 1960 when opposition to the planting of exotic species began. However, in 1985 policies changed again with the Conservation Reserve Program and between 1985 and 1987, 7.5 million ha were sown with this plant species (Pyke 1990). *A. desertorum* is now widely spread in this area.

Although small and numerous seeds characterize many invasive plant species, the tradeoffs between plant longevity, seed size, dispersal, and seed longevity, mean that no strong relationship exists between invasibility and seed characteristics. The presence of a seed bank can benefit the persistence of a species, but phenotypic plasticity of flowering times can also buffer the impacts of environmental heterogeneity.

Evaluating seed banks is tedious. It is possible to measure dormancy by collecting seeds and observing germination following the application of appropriate conditions such as water, covering with soil or scarification. It is also possible to measure the impact of predators on seeds in the soil experimentally by placing seeds out in containers with different mesh sizes that can differentiate among potential predators: mammals, birds, beetles, ants, etc. Seed banks in the field can also be evaluated by removing any flowering plants and monitoring seedling germination over time. An effective way of estimating seed banks is by collecting soil samples to a specific depth and either sifting out seeds or moistening the soil and counting seedlings. However, the counts from these measurements are quite variable. The existence of a seed bank can make eradication of plants impossible and the impact of seed predators as control agents ineffective. Seed banks certainly have strong impacts on the persistence of invasive plants and should be evaluated in population studies of species being considered for biological control programs.

Vegetative reproduction

Vegetative reproduction causes the dynamics of plant populations to differ from those of most animals and allows plants to respond to environmental conditions in a variety of ways. It also can make it difficult to define an individual plant. The frequency of vegetative reproduction varies among plant groups and among geographic locations. Plants with vegetative reproduction are more frequent in stressful environments: alpine areas, high latitudes, herbs in boreal forests, ecosystems influenced by fire and grazing, and aquatic environments (Abrahamson 1980). Vegetative reproduction allows a single individual to colonize an area once it is established, e.g. Japanese knotweed in Britain (Chapter 2).

The advantages of vegetative reproduction can be related to the stability of the habitat and the density of the population. At low population density clonal reproduction allows the establishment and spread of individuals. However, when plant density is high, seed production and dispersal are advantageous for finding new and perhaps more suitable sites. These predictions hold for situations in which the habitat is relatively stable, but they might not hold when the habitat is unstable and unpredictable. Here no clear pattern of switching from vegetative to seed reproduction is expected (Williams 1975, Abrahamson 1980, Figure 4.12).

Tradeoffs between vegetative and seed reproduction might also occur with succession in a habitat (Chapter 3, Abrahamson 1980,

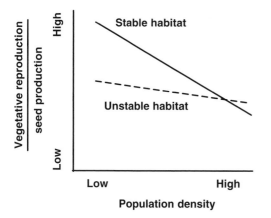

Figure 4.12. In a stable habitat vegetative reproduction will be favored over seed production at low density when space is not at a premium, but seed production will be favored at high density when dispersal to new sites might be favored.

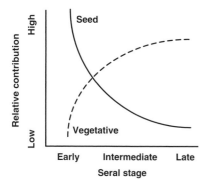

Figure 4.13. In the early stages of succession seed production will be favored over vegetative reproduction. In the late stages of succession, when most patches are utilized, vegetative reproduction can promote persistence. After Abrahamson (1980).

Figure 4.13). In early stages of succession, rapid growth and reproduction are advantageous characteristics of ruderals in Grime's system. As space becomes more limiting, vegetative reproduction may be advantageous for persistence. In unstable habitats, early stages of succession will frequently occur following disturbance, and here communities are expected to have lower ratios of vegetative to seed production than in stable habitats (Figure 4.12). Abrahamson (1980) reviews case studies of plant characteristics in different stages of succession. Studies of invasive plant species

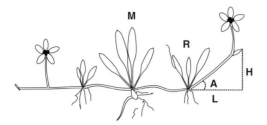

Figure 4.14. Schematic illustration of a *Ranunclus reptans* plant. M is the mother ramet, R is a rosette. Changes in growth form with density or exposure can be measured by the angle of the first internode and the stolon height. After van Kleunen (2001).

provide an excellent arena for further work in this area and are discussed later in this chapter.

The growth form of colonial plants falls into two categories. Compact, aggregated ramets have been described as a 'phalanx' (Lovett Doust 1981) and this growth form characterizes many tussock grasses. This growth form may be beneficial in competitive environments because it resists invasion by other plants. Plants with more loosely aggregated ramets are classified as having a guerrilla growth form. This form facilitates spread into more open spaces in a heterogeneous habitat with low vegetation cover (van Kleunen 2001).

One of the most extensively studied examples of vegetative spread is the clonal herb *Ranunculus reptans* (Figure 4.14). This successful species ranges throughout the temperate and boreal–subarctic zones of Europe, Asia and North America. It colonizes disturbed areas such as the varying shorelines of lakes and rivers. *Ranunculus* reproduces both vegetatively, through stolon growth, and sexually, through flowers arising from rooted rosettes that occur along the stolon. In open areas, *Ranunculus* stolons grow horizontally, along the ground, and in sites with other vegetation stolons grow more vertically, at an angle from the ground (Figure 4.14). In greenhouse experiments, van Kleunen and Fischer (2001) compared the growth patterns of *Ranunculus* clones from open areas and areas in which plants experienced competition from grasses. Clones were grown in experimental trays in which one half was bare soil and in the other the grass *Agrostis stolonifera* was established. The clones from sites in which competition from other vegetation occurred naturally responded more strongly to being grown in a competitive environment by increasing the vertical angle and the length of the first stolon internode. This growth form increases exposure of the leaves to the sun in an environment where

Table 4.6. *Contrasting responses to density of clonal plants showing guerrilla, spreading growth form and phalanx, compact growth form*

Density	Spreading growth form	Compact growth form
Low	Vegetative	Sexual – seeds
High	Sexual – seeds	Vegetative

shading by other plants is a factor. These experiments showed a genetic component to the plasticity of growth form of *Ranunculus*. More plastic genotypes adopted different growth forms in bare patches and vegetated patches, and produced more flowers and rosettes than the less plastic genotypes.

Van Kleunen *et al.* (2001) also studied the response of *Ranunculus* to intraspecific competition by comparing the sexual and vegetative allocations of rosettes planted at high (1466 rosettes m^{-2}) and low (293 rosettes m^{-2}) densities in trays in the glasshouse. After 119 days clones in the low-density treatment produced 155% more rosettes and 227% more rooted rosettes than clones in the high-density treatment. The number of rosettes in the high-density treatment remained higher than in the low-density treatment. The proportion of flowering rosettes was initially higher for the high-density clones but, by 111 days, this and the ratio of flowering rosettes to rooted rosettes were the same between the two treatments. The seeds from the high-density plants tended to be larger than those from the low-density plants, and the germination percentage was significantly higher for the seeds from the high-density treatment. Initially (by day 75) the size of rooted rosettes was greater in the low-density treatment but this difference disappeared by day 115. However, the variation in size of the rooted rosettes was higher by day 115 in the low-density plots.

These experiments indicate an increased allocation to sexual reproduction by *Ranunculus* at high population densities. This response might be expected to differ between clonal plants that have a spreading, guerrilla type growth form such as that of *Ranunculus*, and those with a compact, phalanx growth form (Table 4.6). A survey of the literature by van Kleunen *et al.* (2001) indicates that this pattern generally holds among 13 clonal plants. Four species with phalanx growth form have reduced sexual reproduction with increased density. Three of eight species with guerrilla growth form have increased sexual reproduction with increased density, two have reduced sexual reproduction and three have no difference. High density did not increase the variation in plant sizes as may be expected with intraspecific competition. This is perhaps because mature clonal plants provide support to young rosettes and thus reduce size variation (Hara *et al.* 1993). In non-clonal plants competitive interactions would be expected to enhance variation in plant size at high density. This difference between clonal and non-clonal plants could influence the dynamics of spread of species associated with their growth forms.

An extreme example of an invasive weed that has spread vegetatively is Japanese knotweed, a perennial species with a widespread distribution in the British Isles where it occurs in 1800 10-km squares out of a possible 3500 (Chapter 2). Part of the success of this species is associated with its extensive woody rhizome system with buds just below the soil surface. Although it produces seeds, no seedlings have ever been observed in Britain. All of the plants are thought to be the offspring of a single, female plant (Beerling *et al.* 1994).

Case study – *Phragmites australis* – a story of successful vegetative reproduction

The common reed, *Phragmites australis*, is a cosmopolitan species that currently occurs in Europe, Asia, Africa, America, and Australia (Tewksbury *et al.* 2002). It is native to Europe where reed beds are important for many migrating bird species. European *Phragmites* populations are currently declining and this is causing great concern. Ironically, in North America, particularly on the Atlantic Coast, *Phragmites* is increasing dramatically in both fresh and brackish wetlands, and monocultures there are associated with declines in waterbirds and wetland wildlife. Whether *Phragmites* is native or introduced to North America remains a mystery. Evidence from ground sloth dung indicates it was present in North America as long as 40 000 years ago and fossil *Phragmites* seeds found in peat samples are 3500 years old. However, the rapid expansion of *Phragmites* in recent years and a paucity of native herbivores feeding on these expanding populations suggests a new introduction.

Phragmites spreads primarily through the growth of an extensive underground rhizome system that makes up approximately two thirds of the plant biomass. The vegetative growth pattern results in homogenous clones with up to 200 culms m^{-2} and plants as tall as 4 m. Vegetative propagation occurs through dispersal of rhizome fragments on equipment, with water flow and by animals. Sexual reproduction can also occur and plants are wind pollinated and self-incompatible. Seeds can be carried by wind and water and may be a component of spread to new sites. However, recruitment from seeds is thought to be low (Tewksbury *et al.* 2002).

The growth dynamics of *P. australis* plants were investigated with a diffusion model by Hara *et al.* (1993). Small young plants are supported by large old shoots, which results in little variability in shoot size. A major difference between clonal and non-clonal plants is that physiological integration of clonal plants reduces size variability among individuals at high plant densities. Therefore, instead of a pattern of winners and losers, at high density common reeds, like other clonal plants, form uniform stands.

Do life history characteristics predict invasiveness?

A variety of characteristics or suites of characteristics are likely to be associated with plant species that are successful invaders. Baker (1974) proposed that the following predisposing characteristics might act either alone or together to increase the probability of a plant being invasive. These include (1) rapid initial growth rate and expansion of the root system, (2) ability to interfere with the growth of neighboring plants, (3) high seed output under optimum conditions, with some seed produced under unfavorable conditions, (4) morphological and/or physiological similarity to native species, and (5) breeding systems which allow self-pollination and outcrossing. Gilpin (1990) suggests that because there is no single description of successful invaders, a statistical characterization of invasive species is required. This means that good databases of species are required with which to make comparisons.

Williamson and Fitter (1996b) compared 26 quantitative and qualitative characters of native and invasive species in the British flora and found 12 distinguishing characteristics that were mostly morphological characteristics. Of the life history traits, only the age at first flowering differed significantly between native and invasive species.

If there are different strategies by which invasive plants succeed, we might expect different suites of characteristics to occur in invasive plants. How plants disperse might vary among invasive plant species. Human activity is of paramount importance to the spread of alien species. Hodkinson and Thompson (1997) examined the ecological traits of British plants commonly dispersed by: (1) soil carried on motor vehicles, (2) topsoil, (3) sugar factory topsoil, (4) horticultural stock, and (5) garden throw-outs. They found two clearly defined groups of species. Plant species spread by the movement of soil had numerous, small and persistent seeds. Plants that were spread by garden throw-outs tended to be tall, spreading perennials with transient seed banks. Thus there are two groups of successful, alien plant invaders; tall, spreading competitors and small, short-lived, fast-growing species with high reproductive outputs.

The primary and associated characteristics of invasive plants as summarized by Rejmánek (1996) are outlined in Figure 4.15. In addition to these another characteristic is fitness homeostasis or the ability of individuals or populations to maintain constant fitness over a range of environments. For individuals this could mean having a 'general purpose genotype' (Baker 1974) or for populations it could involve maintaining genetic variability. An example is shown in Figure 4.16 in which the physiological performance of eight European grass species over

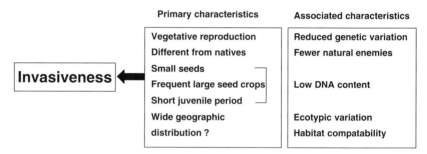

Figure 4.15. Primary characteristics of invasive plants outlined by Rejmánek (1996).

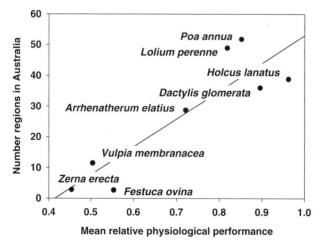

Figure 4.16. Relationship between the distributions of eight introduced, European grass species in Australia and their relative physiological performance over 15 nutrient concentrations. From Rejmánek (2000).

15 nutrient concentrations was compared with their distributions in Australia. Species with high relative physiological performance had broader geographical distributions (Rejmánek 2000).

Considerable variation in life history characteristics exists among species that have been successful invaders and therefore the patterns are complex. For example, Gerlach (2001) points out that Lonsdale (1994) claimed that seed weight was not a correlate of invasiveness in herbaceous and woody legumes and grasses introduced to northern Australia and this contradicts the findings of Rejmánek (2000) for invasive plants elsewhere. However, legumes typically produce heavier seeds than grasses and differences in the areas of origin of the introduced species – temperate, subtropical, and tropical – were confounded in this comparison.

The studies that have searched for distinguishing characteristics of invasive and non-invasive plants have not always agreed. As studies of invasive species and their characteristics accumulate, however, it will possible to refine the search for consistent characteristics of invasive species. Kolar and Lodge (2001) summarized studies to date. Characteristics associated with becoming invasive after establishment include: the region of origin, the family or genus of the invasive, the history of invasion by other relatives (species, genus or family), vegetative reproduction, low variability of seed crop and a short juvenile period. Characteristics of invasives that varied widely in importance were region of origin and dispersal mechanisms.

In many cases the importance of characteristics of plant species to their invasiveness potential will be habitat dependent. Given this, teasing out patterns will be challenging as will the eventual task of incorporating invasiveness with the pest status of the plant (Rejmánek and Reichard 2001). Continued research is both necessary and crucial for the development and testing of patterns and relationships. This may best be done by concentrating on particular plant groups such as trees (Box 4.1).

Predicting invasive species and the design of quarantine regulations

The strongest pressure for defining the characteristics of invasive plant species, is the need to design regulations that will prevent the introduction of problem species. Schemes for assessing plant species prior to deliberate introduction have been proposed by Panetta (1993), Reichard and Hamilton (1997), and White and Schwarz (1998). White and Schwarz (1998) list traits that are common to plants desired by

---

Box 4.1 · *Life histories and forestry trees as invaders.*

A group of plants for which the life history characteristics are well known are trees that have been planted widely for commercial forestry and agroforestry. Some of these have caused major problems in their new habitats. *Pinus* spp. are especially problematic with 19 species having become invasive over large areas of the southern hemisphere. The most invasive of these species have a predictable set of life history attributes: low seed mass, short juvenile period, and short interval between large seed crops. Richardson and Higgins (1998) suggest that the extent of invasions of trees planted in forest projects can be explained by information on species attributes, residence time, the extent of planting, ground cover characteristics, latitude, disturbance, and the resident biota.

Acacia species have also been widely used for forestry and, like the pines, have become invasive in many areas. *Eucalyptus* species occur on many lists of weeds, but even though they produce large quantities of seeds, they have been less invasive from forestry plantings than have pines and *Acacia* spp. The reasons for this are not known.

Tree species can be screened for invasive potential by preplanting tests. For example, Aronson *et al.* (1992) screened 40 legume species and were able to exclude some species with high potential for invasiveness. Sterilizing exotic trees planted in forestry projects may be possible in the future, but amelioration of current problems is only feasible if a 'polluter pays' policy is established to fund habitat reclamation. Preventing the spread of commercially important trees requires careful evaluation of risks, costs, and benefits and the political will is often lacking to curb the introduction and planting of exotic trees for forestry. These trees have the life history traits that characterize early-successional plants and these same traits also make them potential invaders (Richardson 1998).

---

gardeners and those of invasive weeds: rapid growth, early and many flowers, prolific seed production, good seed germination, efficient dispersal and no major pests. Reichard and Campbell (1996) developed a scheme for catagorizing the invasiveness of plants by comparing the traits of 235 species of plants known to be invaders with those with 87 non-invasive plants. White and Schwarz (1998) tested the success of this scheme to predict the invasiveness of known plant invaders and found

Table 4.7. *Evaluation scheme proposed by Panetta (1993) to predict invasiveness of plants*

Criterion	Point value
Is the species free-floating aquatic or can it survive and reproduce as a free-floating aquatic?	20
Is it a weed elsewhere?	20
Are there close relatives with a history of invasion to similar habitats?	10
Is it spiny?	10
Are diaspores spiny?	10
Is the species harmful to animals?	8
Does the plant produce stolons?	5
Does the plant reproduce vegetatively?	8
Are diaspores wind-dispersed?	8
Are diaspores dispersed by mammals or machinery?	8
Are diaspores dispersed by water?	5
Are diaspores dispersed by birds?	5

that 85% would have been rejected by the scheme, 13% would have been held for further examination and 2% would have been accepted.

Panetta (1993) proposed a scheme based on earlier work of Hazard (1988) (Table 4.7). This system rejects aquatic plants outright, which causes potential friction with the majority of those selling aquatic plants for aquaria and ponds.

A similar but more elaborate scheme was proposed by Pheloung *et al.* (1999) (see Box 4.2 for questions). White and Schwarz (1998) tested the ability of this system to reject plant species already known to be invasive in Australia and of the current invaders, 84% would be rejected by this scheme, 16% fell into the category of requiring future study and none would have been accepted. Pheloung *et al.* (1999) successfully adopted their original model to fit conditions in New Zealand.

Finally, a recent report for the US National Academy of Sciences (NAS 2002a) lists six indications of the potential invasiveness of plants as: (1) history of invasiveness elsewhere, (2) occurs with disturbance, (3) has means of rapid and efficient dispersal, (4) has fleshy fruit, (5) has high reproductive output, and (6) is closely related to native flora. They conclude that critical data are insufficient to develop broad principles or procedures for identifying the invasive potential of plants, but that conceptual bases for understanding invasions and expert judgments should be transformed into 'transparent, repeatable, quantitative, and comprehensive predictions'.

Box 4.2 · *Questions forming the basis of the Weed Risk Assessment model.*

Domestication/cultivation
1.01 Is the species highly domesticated?
1.02 Has the species become naturalized where grown?
1.03 Does the species have weedy races?

Climate and distribution
2.01 Is the species suited to the climate?
2.02 Quality of the climate match data?
2.03 Broad climate suitability (environmental versatility)
2.04 Native or naturalized in regions with extended dry periods
2.05 Does the species have a history of repeated introductions outside its range?

Weed elsewhere
3.01 Naturalized beyond the native range
3.02 Is it a garden/amenity/disturbance weed?
3.03 Is it a weed of agriculture/horticulture/forestry?
3.04 Is it an environmental weed?
3.05 Congeneric weed

Undesirable traits
4.01 Produces spines, thorns or burrs
4.02 Allelopathic
4.03 Parasitic
4.04 Unpalatable to grazing animals
4.05 Toxic to animals
4.06 Host for recognized pests and pathogens
4.07 Causes allergies or is otherwise toxic to humans
4.08 Creates a fire hazard in natural ecosystems
4.09 Is a shade-tolerant plant at some stage of its life cycle
4.10 Grows on infertile soils
4.11 Climbing or smothering growth habit
4.12 Forms dense thickets

Plant type
5.01 Aquatic
5.02 Grass

5.03 Nitrogen fixing woody plant
5.04 Geophyte

Reproduction
6.01 Evidence of substantial reproductive failure in native habitat
6.02 Produces viable seeds
6.03 Hybridizes naturally
6.04 Self-fertilization
6.05 Requires specialist pollinators
6.06 Reproduction by vegetative propagation
6.07 Minimum generative time

Dispersal mechanisms
7.01 Propagules likely to be dispersed unintentionally
7.02 Propagules dispersed intentionally by people
7.03 Propagules likely to disperse as a produce contaminant
7.04 Propagules adapted to wind dispersal
7.05 Propagules buoyant
7.06 Propagules bird-dispersed
7.07 Propagules dispersed by other animals (externally)
7.08 Propagules dispersed by other animals (internally)

Persistence attributes
8.01 Prolific seed production
8.02 Evidence that a persistent propagule bank is formed
      (>1 year)
8.03 Well controlled by herbicides
8.04 Tolerates or benefits from mutilation, cultivation or fire
8.05 Effective natural enemies present in country

The report of the National Academy of Sciences committee recommends that risk assessments, based on the likelihood of an event and its consequences, are more useful for management than predicting the invasive potential of plant species. Risk assessments, as outlined in Table 4.8, can incorporate uncertainty, but they share the limitations of attempting to predict the invasiveness of plants described above. The process of risk assessment clearly documents decision criteria and these can be further evaluated in the future.

Considerable effort has been directed at attempting to predict which plant species may become invasive. Some systems seem to hold promise

Table 4.8. *An outline of qualitative risk-assessment procedures in which the risk potential is based on a consideration of the likelihood of introduction and the consequences of introduction. Scores given for the different characteristics would be added together to give an assessment score*

Risk potential

LIKELIHOOD OF INTRODUCTION

Frequency of imported material	**Pest opportunity**
Number of ships, etc.	Survive treatment
	Survive shipment
	Escape detection
	Move to suitable environment
	Move to suitable host

CONSEQUENCES OF INTRODUCTION
Climate–host interaction

Host range	**Dispersal potential**
Multiple families	High reproductive potential and rapid dispersal
Multiple species	High reproductive potential or rapid dispersal
One or more species	Neither of the above
Economic impact	**Environmental impact**
Reduced crop yield	Significant direct impacts
Reduced commodity value	Impacts on endangered species
Loss of markets through quarantine	Requires chemical control
	Requires biological control

Adapted from NAS (2002a).

and have been tested with retrospective data. No system will be totally accurate for every plant species. Documentation of the decision making process may be most important. Experiments can be used to gain valuable information of species characteristics, but they are slow and expensive. Applying the cautionary principle to plant introductions or the importation of materials possibly contaminated with plant diseases (see Chapter 6) is the best approach.

Conclusions

Life history traits influence the reproduction and survival of plant species and should be directly related to their ability to establish and invade new areas. These life history characteristics can be under genetic control and respond to selection, or they can vary in response to environmental conditions. While the characteristics that distinguish invasive and non-invasive plant species are not totally consistent, patterns do emerge. Plants with small and numerous seeds, physiological robustness, small genome size, good dispersal ability and lack of specific mutualisms tend to be invasive. For aquatic plants in particular, vegetative reproduction is associated with the spread of introduced plant species. Several schemes have been proposed for screening plants to reduce the continued introduction of species likely to become detrimental weeds. These schemes have been tested and show some promise. However, strong commercial pressures to introduce plant species as ornamentals, aquarium plants and for erosion control act against implementation of strict quarantine regulations.

5 · Population ecology and introduced plants

Why study plant populations?

In this book we have thus far focussed on introduced plant species and their impacts on and relationships to native plant communities. We are interested in the patterns of spread, the stability of established populations, and the impact of control procedures, particularly biological control. For all of these considerations environmental and biological heterogeneity are major perturbing factors. Land managers are confronted with serious questions about interactions among populations of plant species, particularly those between native and introduced species. Therefore it becomes imperative to measure, predict and interpret changes in populations.

When plant species move into new communities their dynamics create challenges for population ecologists. That introduced species are able to establish and invade a new plant community indicates: (1) the presence of empty niches in the native community, (2) that the introduced species itself creates a new niche, or (3) that the introduced species is a superior competitor able to respond to disturbance or utilize resources better than existing species. The introduced plant species will probably also differ from species in the native community by having less herbivore and disease pressure (Chapter 3). During establishment and spread, the introduced species will increase in density and distribution. Following the invasion, the dynamics of the invader may not be very different from those of the native species other than that densities will be higher and therefore intraspecific competition will be strong. If biological control is attempted, a new selection pressure will be placed on the plant, and an experiment in population ecology will unfold – a test of the role of herbivores in plant population dynamics. All of these considerations require an understanding of plant population dynamics.

Table 5.1. *Potentially limiting or regulating factors for plant populations*

Limiting factors – set average density	Regulating factors – respond to density
Soil characteristics	Intraspecific competition
Nutrients	Herbivores that respond to plant density
Moisture	Diseases that respond to plant density
Texture	
Disturbance	
Climate – temperature and precipitation	
Sun exposure	
Interspecific competition	
Herbivores not responding to plant density	

What determines plant population densities?

It is useful to think of the processes influencing plant populations in two categories: **limiting factors** determine the average density of a plant in a particular area and perhaps its geographic distribution, and **regulating factors** increase in impact as plant density increases and diminish in impact when density declines. Both of these are variable and they can interact in ways that make it difficult to predict future plant densities. Some limiting and regulating factors are listed in Table 5.1.

Plant densities are influenced by factors associated with the soil in which they grow and by the organisms with which they interact. The densities of plants can also affect other organisms. These complex interactions can be looked at as top-down trophic cascades or and bottom-up influences (see Box 5.1). Demonstrating whether herbivores drive the dynamics of plant populations, or if plant populations drive the dynamics of herbivores, is difficult (Myers 1980).

Before thinking about how limiting and regulating factors work, it is important to keep in mind that overall plant population density is determined by four processes: reproduction, immigration, mortality, and emigration. Because dispersal into and out of populations is difficult to measure, population ecologists concentrate on reproduction and mortality. These can be related to plant density either proportionally or absolutely (Figure 5.2). For example, a herbivore that always consumes the same number of plants regardless of the density of plants available will have less of an impact when there are many plants than when there are few. However, a herbivore that increases its intake of plants in direct proportion to the increase in plant availability will have a constant impact on

Box 5.1 · *Top-down and bottom-up influences in the dynamics of plant populations.*

The processes that determine plant density have frequently been cast in the paradigm of top-down or bottom-up influences (Figure 5.1). Top-down influences (trophic cascades) are those that reduce plant density by stressing or removing plants. These are generally the impacts of herbivores or diseases, and they have the potential for responding to the density of the plants. Bottom-up influences affect plant density and/or quality and can modify the populations of herbivores. Soil quality, moisture, disturbance or even competitors could be bottom-up influences that act through the plant to the natural enemies attacking the plant. The observation that the world is green, i.e. the leaves of plants in general show little influence of herbivores, suggests that top-down influences are less important to plants than are bottom-up effects (Hairston *et al.* 1960). If top-down influences

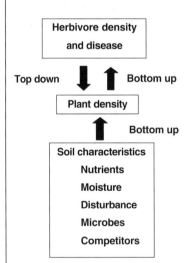

Figure 5.1. Top-down and bottom-up influences on plant densities as envisioned in the trophic cascade hypothesis based on Carpenter *et al.* (1985).

on plant population densities are not strong, then biological weed control would not be expected to be successful. This will be further investigated in Chapter 7.

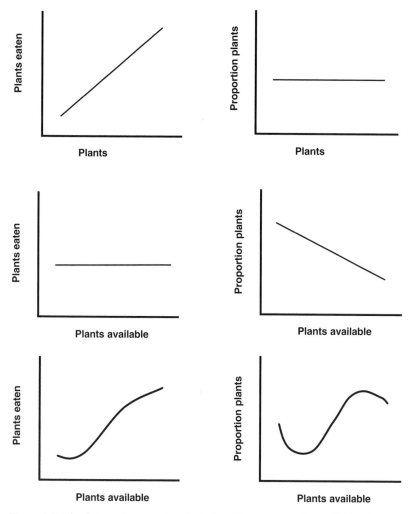

Figure 5.2. Absolute and proportional relationships between the available plants and the number and proportion of plants eaten by herbivores. Only when proportionately more plants are taken with an increase in density can herbivores have a regulating impact on the density of the food plants.

the plant population. A herbivore that increases its impact on plants above the rate of increase in plant availability will have an increasing impact on plant density. Thus it can reduce the host plant density in a regulatory manner. Without taking into consideration the rates of herbivory in relation to plant density, we cannot say what the impacts of herbivores might be. In addition, without knowing if plant populations compensate

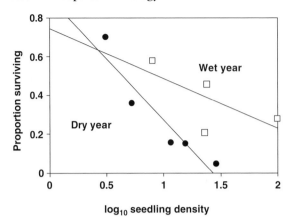

Figure 5.3. Relationships between the proportion of diffuse knapweed (*Centaurea diffusa*) seedlings surviving over a summer and the survival of seedlings in five sites in the interior of British Columbia, measured in a dry year and a wet year. One site was burned between the two years of observation so only four areas were sampled in the wet year. Data from Myers and Risley (2000).

for increased herbivore pressure by better survival of remaining plants, we cannot predict long-term outcomes. How herbivores and disease respond to host density is of particular interest to their effectiveness as biological control agents, as will be discussed in Chapter 7.

While most limiting factors are specific to the site or environment where the plant is growing, regulating factors are largely biotic and involve the interactions between organisms. An exception might be rainfall and soil moisture, for which the impact might change with the density of the plants in such a way that plants at low density would have more soil moisture per individual than those at high density. For example, seedling survival might be higher for sparse populations during a drought than for dense populations during normal rainfall conditions. On the other hand, soil moisture could change the relationship between plant survival and plant density. In Figure 5.3 the relationship is shown between seedling survival and density of diffuse knapweed, *Centaurea diffusa*, at several study sites measured in a wetter and in a drier year. In the dry year, the drop in survival with an increase in density is steeper. The increase in seedling survival with reduced density is a compensatory mechanism that buffers plant populations against changing conditions. The relationship between rainfall and survival means that the density of plants changes with levels of precipitation as shown in Figure 5.4, again with knapweed.

Soil moisture might also determine whether a site is suitable for a particular plant species and this can change over a short distance in uneven

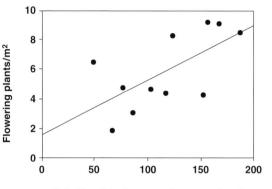

Figure 5.4. The relationship between the density of diffuse knapweed and rainfall over 11 years for a study area near White Lake, British Columbia (J. Myers unpublished).

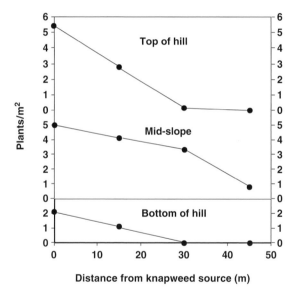

Figure 5.5. Density of flowering diffuse knapweed plants measured along three transects on a grassland hillside at Buse Hill, British Columbia. The invasive species was introduced at this site from vehicles driving along a dirt track running over the hill approximately 50 m from the 0 location on the transects and was spreading along the hillside. Data from Berube and Myers (1982).

terrain. Diffuse knapweed can invade grassland sites in dry areas (Berube and Myers 1982). However, the density and rate of spread of this introduced weed into a rangeland in British Columbia varies with the position on a hillside (Figure 5.5). Soil moisture is slightly higher at the bottom of the hill and here knapweed was less dense than in drier sites at the middle and top of the hill. This relationship between location on the slope and density is probably associated with the knapweed having a reduced competitive advantage with grasses in areas with increased soil moisture. This site is on the wetter end of the range of sites infested by diffuse knapweed. Thus, soil moisture can be both limiting and regulating for plant populations by influencing the average density and the strength of intraspecific competition. Intraspecific competition is clearly a density related process and therefore is regulatory. This is the concept of self-thinning, which has been much discussed in plant ecology.

Self-thinning and the 3/2 rule

Levels of plant growth and reproduction are determined from the amount of energy from the sun and the availability of nutrients and moisture in the soil. For a particular site there will be a limit on the total plant biomass that can be produced, but this can be divided in various ways among individuals and among different species. Plant densities will almost always be high when plants are small and the mass is shared among many individuals. As plants grow, some individuals die and others will be released from competition to grow larger. The relationship between plant density and yield is of particular importance in agriculture and forestry. A farmer will want to plant the number of seeds of a crop that will give the highest yield at the end of the season. Similar relationships confront the silviculturalist who needs to decide how many tree seedlings to plant in a reforestation project.

For sexually reproducing plants, seedling densities will be determined by seed production, available sites, and seed germination. Between the seedling and the reproductive stage plants will grow, some will die and density will decline. The decline in density associated with crowding is known as self-thinning. This is a density-dependent process that is a general phenomenon in plant populations. A number of studies have found a relationship between the mean weight of plants and the density of plants to approximately fit the following:

$$\text{mean weight} = c\,N^{-k}$$
$$\text{or log mean weight} = \log c - k \log N$$

where N is plant density.

For many of these studies the slope of the relationship approximates $k = -3/2$, but whether this is an ecological 'law' has been controversial (White 1981, Weller 1987, Lonsdale 1990, Lawton 1999). More recent considerations have attempted to derive the self-thinning rule from first principles based on the allometry of the total available flux of energy and materials $\simeq$ the mass of plants$^{3/4}$ (Enquist *et al.* 1998, Torres *et al.* 2001).

The important applications of self-thinning to invasive species are that (1) high densities of seedlings will be reduced to much smaller numbers of reproductive plants as the plants grow, and (2) given the limited levels of resources for plant growth in any area, the addition of invasive species will generally reduce the sizes or densities of plants native to the site.

Are plants seed limited?

Many invasive plants are excellent seed producers and this contributes to their local reproduction as well as their potential for dispersal. Even though plants may produce hundreds of seeds, for a sexually monomorphic species only one seed per individual need survive through to the reproductive stage for the population to maintain a stable density. That means, for example, if a plant produces 100 seeds, seedling mortality will be 99% on average in a 'stable' population. The recruitment to plant populations can be limited by two factors: the number of seeds and the number of suitable microsites in which seeds can germinate. Once seedlings are established, further changes in density occur through self-thinning. Crawley (1990a) suggested that in general plant populations are microsite limited. Louda (2001) argues that the importance of seed limitation should not be overlooked. Her ideas are based on her own experiments of reducing seed predators of plants over a density gradient and studying the response of plant populations to seed additions and seed predator reduction. One experimental approach to testing if seeds are microsite or seed limited is to add seeds to an area and determine if plant populations increase. This is the opposite of Louda's approach, in which she removed seed predators to increase seed numbers. In addition, disturbing the soil is a way of testing if increasing microsites also increases plant density and, as discussed in Chapter 4, this is often the case.

To look at this question Eriksson and Ehrlén (1992) studied the responses of 14 plant species in deciduous and coniferous forests of central Sweden to test for seed and/or site limitation. Three species had increased seedling production in undisturbed sites with seed addition, six species had increased seedling production with seed addition and disturbance, and

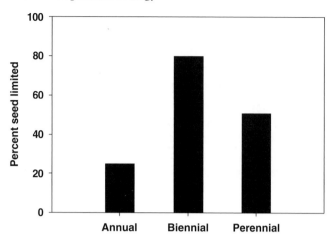

Figure 5.6. Percentage of annual (*n* = 4), biennial (*n* = 10) and perennial (*n* = 47) species that are seed limited. Data from Turnbull *et al.* (2000).

five species had no seed germination whatsoever. The authors interpreted this last category as indicating germination limitation rather than seed limitation. However, drawing conclusions prior to the self-thinning phase of the population growth limits interpretations about the impacts of seed or site limitation on population density. A more extensive review of seed limitation was carried out by Turnbull *et al.* (2000). They considered 27 studies involving 90 plant species in which seed augmentations had been done. Of greatest interest are the 64 species for which establishment was monitored for over a year. In 34 (53%) of these, plant establishment was increased following seed sowing. Of the experiments for which increased seed establishment was observed for 12 months or more, few annual plant species showed seed limitation (Figure 5.6) and few of those in grasslands were seed limited (Figure 5.7).

From these studies we can conclude that seed limitation, as indicated by an increased number of plants for at least a year following seed augmentation, occurs in approximately half of the species studied. More long-term studies are required to determine the contribution of seed augmentation to reproductive plant density.

Numerous models of biological control have attempted to show whether a reduction in seeds would reduce plant density (see Chapter 8). In almost all of these studies, the projections are that 95 to 99.9% of seeds would have to be destroyed to reduce plant density (Smith *et al.* 1984, Noble and Weiss 1989, Cloutier and Watson 1990, Powell 1990b,

Figure 5.7. Percentage of plants from grasslands ($n = 26$), dunes ($n = 4$), early successional habitats ($n = 9$), and woodlands ($n = 18$) that are seed limited. Data from Turnbull *et al.* (2000).

Hoffmann and Moran 1998, Myers and Risley 2000, Parker 2000). Paynter *et al.* (1996) predicted that seed predators would not be successful in limiting broom population densities in grassland sites of New Zealand. Invasive plants appear to have evolved under conditions that select for resilience to the attack of seed predators. This has vital implications for biological control programs.

Demographic parameters

As mentioned above, population change from one generation to the next depends on births (B) − deaths (D) + immigrants (I) − emigrants (E). Population density in one year is designated as N_t and that in the next as N_{t+1}. If these parameters can be measured the change in population density can be measured as $N_{t+1} = N_t + B - D + I - E$. This relationship is frequently simplified by assuming that immigration and emigration are equal and cancel each other out. This will not be the case for an invading plant that is moving into areas still lacking the new species. Here we consider the case in which a plant has become established at a site. This relationship is complicated because plants tend to survive better at low density (Figure 5.3) and to grow larger and produce more flowers and seeds when they are less dense (Figure 5.8). Therefore, although both B and D can be defined as B/N_t and D/N_t, because the relationships between density and births and deaths change with environmental conditions, these are certainly not constants for any population.

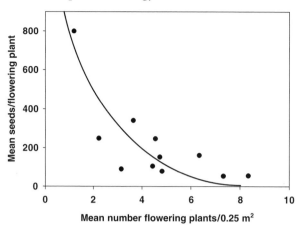

Figure 5.8. The relationship between density and flower production for diffuse knapweed in the dry interior of British Columbia. After Myers and Risley (2000).

For an annual plant species the change in population density from one generation to the next can be determined from the ratio of the population density between years (N_{t+1}/N_t). This value is known as the net reproductive rate, R_0. If the population is increasing, as might be the case for an invading weed, R_0 will be greater than 1 and the population (N_t) can be multiplied by R_0 to determine N_{t+1}. An equation based on exponential increase can be used to determine the population in the future. The expression for this is $N_{t+x} = N_t e^{rx}$ in which e is the base of natural logarithms and r is the intrinsic rate of population increase, the maximum rate of growth in the absence of resource limitation or competition in the given environmental conditions and x is the number of years. The finite rate of increase of the population, λ is equal to e^r and therefore $r = \ln \lambda$. Because births and deaths are influenced by a variety of ecological factors, estimating r or λ and thus predicting future populations is unlikely to be very accurate, with the exception of the general conclusion that for established populations and in the absence of succession, λ on average will be 1.

Many weeds are short-lived perennials; for these, to determine λ one must take into consideration the generation time τ, the time between reproduction by the mother plant and that of her daughter. In this situation, the net reproductive rate R_0 is equal to $N_{t+\tau}/N_t$. R_0 is measured over the generation time and thus the finite rate of increase for a perennial species is $\lambda = R_0^{1/\tau}$ and the instantaneous rate $r = \log_e(R_0)/\tau$. Instantaneous rates are often used because they are more simple to deal with mathematically. They can be thought of as the change in the population over a number of tiny time periods. When low, finite and instantaneous rates of increase yield similar results.

These descriptions of plant population dynamics are unrealistically simple. For some species, even estimating the generation time can be complicated. For example, diffuse knapweed is described as a short-lived perennial plant but detailed studies by Powell (1990b) showed that the rosette stage can be as short as one summer or as long as 5 years. The survival and length of the rosette stage is strongly influenced by plant density. In Figure 5.9A, R_0 is plotted on the density of flowering knapweed plants without consideration for the generation time and in Figure 5.9B the density of flowering plants in year N_{t+1} is plotted on the density of flowering plants in year N_t. These figures show that at high densities R_0 is less than or close to 1, but at low density R_0 is 2 and above. During Powell's study the density of flowering knapweed plants varied over a $3\times$ range. Part of this variation in population density was related to rainfall (Figure 5.4) and this, plus the relationship of seedling survival (Figure 5.3) and seed production to density (Figure 5.8), generates variation in density from year to year.

Monitoring populations

Field studies of plant populations require adequate design and sufficient sample sizes. Some basic considerations of field sampling are outlined in the Appendix. Choosing a sampling technique that is efficient in terms of time and money is an important consideration. It must also yield a representative sample and one that is sufficiently large to have statistical power. For many this is the fun part of population biology. Actually being in the field can yield important insights that must not be lost in the complexities of data analysis. Monitoring should be done with particular questions in mind. The most important step in a population study comes at the very beginning and that is to ask the question, 'will this monitoring procedure yield the data required to answer my research question?' These data can be used to estimate the demographic parameters described above and can be the basis for population models. Field data can also be used to describe reproduction and survival as described below. Good field data, particularly long-term data, are a valuable resource.

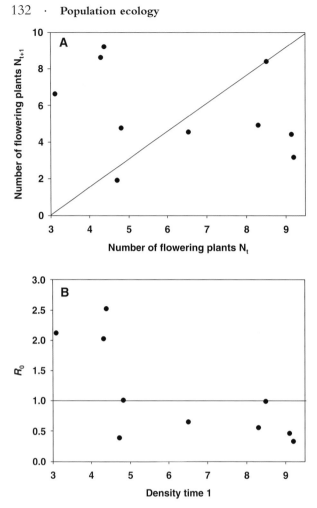

Figure 5.9. (A) R_0 (net reproductive rate) plotted on the density of flowering knapweed plants at White Lake in British Columbia, and (B) the number of flowering knapweed plants in year N_{t+1} plotted on the density in year N_t. Data from J. Myers and C. Risley (unpublished).

Life tables and key factor analysis

Life tables summarize information on the survival and reproduction of individuals in populations. They can be based on tracking individual plants through their lives or by estimating survival through the change in density of progressive life history stages. Table 5.2 shows a life table over one season for diffuse knapweed. Diffuse knapweed is a short-lived perennial and the time for development depends on conditions of crowding and

Table 5.2. *Life table for diffuse knapweed during one summer*

Stage	Number alive N_x	Number dying d_x	Proportion dying q_x	$\log_{10}$ no. alive	$\log_{10} N_x - \log_{10} N_{x+1}$ k-value
Seeds	1043	839	.80	3.02	0.71
Viable seeds	204	175	.86	2.31	0.85
Seedlings	29	16	.55	1.46	0.35
Rosettes	13	3	.23	1.11	0.11
Killed by beetles	10	2	.20	1.00	0.10
Flowering plants	8	0	.00	0.90	
Potential seeds	944	944	.39	2.97	
Seeds reduced by flies	354	550		2.55	0.42
Total seeds	354				$K = 2.54$

Data from C. Risley and J. Myers (unpublished).

precipitation. This life table begins with the seeds and carries through the development stages to seeds in the next generation. Seeds in the soil were estimated by sorting through soil cores. Not all of the seeds are viable and by germinating the seeds it was possible to estimate this first cause of mortality. Subsequent life stages were counted in quadrats in the study area. The number alive, N_x, is given for each life stage. The number dying at each age interval d_x is determined by subtracting the number alive at the next interval from the number alive at the previous interval. The amount of mortality over age intervals can be determined by summing the d_x values and the intensity of mortality at different intervals is indicated by the mortality rate, q_x, determined by dividing the number dying by the number alive at the beginning of the interval. The number dying over life stages can be added but the q_x values cannot be summed.

Varley and Gradwell (1960) proposed using k-values for measuring the rate of mortality in life table studies. The k-values are derived by subtracting the $\log_{10}$ of the number of individuals still alive in subsequent intervals. An advantage of k-values is that they indicate the intensity of mortality at each stage and they can also be summed over the study to give the total K for the generation mortality. This latter quality is useful if one is studying populations over several generations in which fluctuations in population numbers and variation in mortality causes and patterns are of interest. This approach is referred to as a key factor analysis. Key factor analysis has been used in studies of animal populations, particularly insects (Krebs 1999, 2001), but has not been widely used in plant studies. However, it could be useful if one had a measure of the impact of herbivores on plant mortality at different stages that could be incorporated into the analysis. Key factor analysis might also be used to determine the stage of the plant at which the greatest amount of mortality occurs and how this is related to the change in density of plants over time. The greatest amount of mortality is not necessarily related to the change in density of the population over time.

We have shown data on diffuse knapweed for only one year here but several years' data are necessary to determine the key factor of population change. In the life table, the greatest loss is associated with non-viability of seeds and the reduction of seed production caused by seed-gall forming flies, *Urophora affinis* and *U. quadrifasciata*. Most of the mortality of rosettes was associated with attack on the roots by the beetle, *Sphenoptera jugoslavica*.

Life tables can also be constructed for longer-lived plants, and in Table 5.3 a life table for the shrub *Acacia suaveolens* in southwest Australia is presented. Initially 1000 plants were marked and these were followed through 17 years. This life table includes not only the survival of plants between years, but also the average seed production of plants as they aged. From these data two curves related to age can be plotted, survival and seed production (Figure 5.10A, B). The survival curve declines gradually with age in this plant and seed production begins for plants in the second year, remains high for several years and then declines between years 4 and 6. The relationship between age, survival and fecundity can be used to determine the reproductive value for each age category. This is the average reproductive contribution an individual of a certain age will make to future generations. It can be seen from Figure 5.10C that in the first year the reproductive value is high for *A. suaveolens* even though individuals have not begun to reproduce. First year plants have high reproductive potential at this stage and good survival, and thus the potential total of the average number of offspring produced over all later stages is high. By the next year the reproductive value is higher because individuals will have actually reproduced then.

The optimum age of reproductive maturity is reached when $\Sigma l_x m_x$ is at its highest value and further delay in reproduction would cause this value to decline. For populations of invasive plants that are not in equilibrium but increasing and expanding into new areas, there may be a premium on reproducing early. The residual reproductive value $(V_x - m_x)$, or the expectation of further reproduction can be adjusted for

Table 5.3. *Life table and fecundity schedule for* Acacia suaveolens, *in Australia*

Age x	Number alive N_x	Survival l_x (N_x/N_0)	Per capita mortality $q_x =$ d_x/N_x	Seeds/ plant m_x	$\Sigma l_x m_x$	$\sum_{i=1}^{i=\infty} \frac{l_{x+i}}{l_x}$	$V_x = m_x +$ $\sum_{i=1}^{i=\infty} \frac{l_{x+i}}{l_x} m_{x+i}$
0	1000	1	0.174	0	97.78	169.25	169.25
1	826	0.83	0.176	41	97.78	135.38	176.38
2	681	0.68	0.233	33	63.92	108.17	141.18
3	522	0.52	0.234	31	41.44	84.41	115.41
4	400	0.4	0.233	31	25.26	60.66	91.66
5	307	0.31	0.248	18	12.86	46.84	64.84
6	231	0.23	0.247	9	7.34	40.07	49.07
7	174	0.17	0.247	9	5.26	33.29	42.29
8	131	0.13	0.115	9	3.69	26.52	35.52
9	116	0.12	0.112	7	2.51	20.32	27.32
10	103	0.10	0.117	5	1.70	15.88	20.88
11	91	0.09	0.121	3	1.18	13.23	16.23
12	80	0.08	0.113	6	0.91	7.95	13.95
13	71	0.07	0.127	–	0.43	7.95	7.95
14	62	0.062	0.113	–	0.43	7.95	7.95
15	55	0.055	0.127	2	0.43	6.18	8.18
16	48	0.048	0.104	4	0.32	2.69	6.69
17	43	0.043	–	3	0.13	0	3

After Silvertown and Lovett Doust (1995) and data from T. Auld and D. Morrison.

whether the population is increasing or declining by multiplying $(m_x + \sum_{i=1}^{i=\infty} \frac{l_{x+i}}{l_x} m_{x+i})$ by (N_t/N_{t+1}). For increasing populations $N_{t+1} > N_t$, and the value of one progeny is less as a proportion of the total population. This means that in increasing populations the residual reproductive value is reduced. The reproductive value is important in the evolution of the life history characteristics of a population and earlier reproduction should be advantageous in increasing populations.

Whether the population is established or spreading can influence survival, reproduction and the reproductive potential of plants. In the Pacific Northwest, Scotch broom, *Cytisus scoparius*, a native of Europe, has become a major weedy shrub in many locations. Parker (2000) studied populations in two habitats, prairie and urban sites, and within those sites in areas in the center of the population and areas on the edge of the spread. In the life table Parker distinguished six life stages: seedlings,

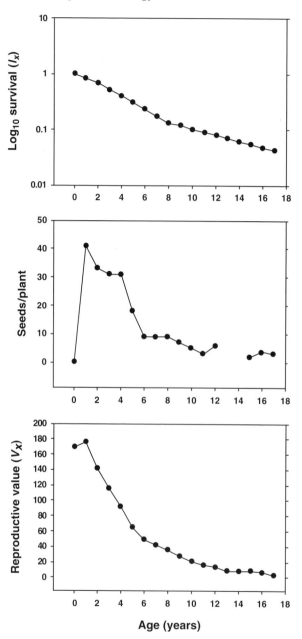

Figure 5.10. Log₁₀ survival, seeds per plant and reproductive value for age classes 0–17 of *Acacia suaveolens*, in Australia. Data presented in Table 5.3.

Table 5.4. *Stage-specific survival, fecundity and reproductive potential of center and edge sites of a population of the European invasive, Scotch broom, at a prairie site in Washington*

	Edge site				Center site			
Life stage	Survival l_x	Seeds/ plant m_x	$l_x m_x$	V_x	Survival l_x	Seeds/ plant m_x	$l_x m_x$	V_x
Seedling	.33	0	0	2284	.07	0	0	1836
Juvenile	.92	19	16	2250	.50	0	0	1836
Sm. Adult	.88	161	142	2238	.66	0	0	1836
Med. Adult	.94	461	433	2046	.76	118	90	1818
Lg. Adult	.95	611	580	1578	.92	483	444	1598
XLg. Adult	.84	1094	919	1094	.97	1058	1026	1058

Data from Parker (2000).

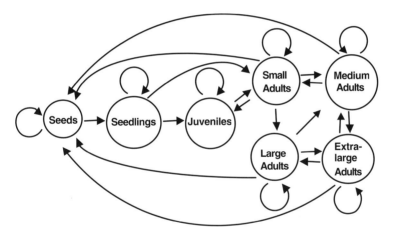

Figure 5.11. Life cycle of Scotch broom (*Cytisus scoparius*), showing possible transitions. After Parker (2000).

juveniles, small, medium, large and extra large adults, and she recorded the survival of the six life stages over a year (Table 5.4). Although only six stages were distinguished, the transitions among these are complex as diagramed in Figures 5.11 and 5.12.

Plants on the edge of the distribution survived better, produced more seeds and began to reproduce earlier than those in the center. Although

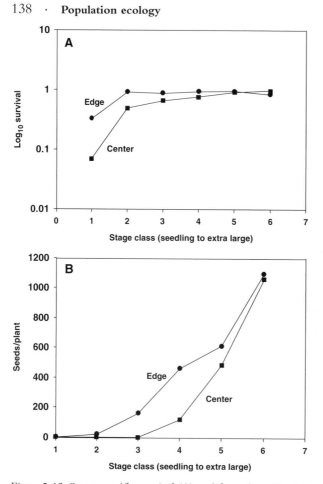

Figure 5.12. Stage-specific survival (A) and fecundity (B) of Scotch broom at the edge and in the center of a population at a prairie site in Washington. Data from Parker (2000).

the density of plants did not differ between the center and edge sites at this location, 5.8 vs. 6.1 plants m^{-2}, the density of large plants differed, 1.50 (center) vs. 0.17 (edge) plants m^{-2}. It is likely that the delayed reproduction of the plants in the center area was caused by environmental conditions rather than selection on the difference in the reproductive potential between the two areas. Theoretically the net reproductive rates (R_0) of these two populations can be determined by summing the $l_x m_x$ values in Table 4.4 and correcting for seedling establishment the following year, as given in Parker (2000), to complete the cycle to the next generation. This gives $R_0 = 2346 \times 0.03 = 70$ for the edge population and $R_0 = 1560 \times 0.0086 = 13$ for the center population, a value that indicates a considerable potential for population increase. The populations of broom did increase between the two years of this study, but not in a way that would have been predicted from the life table. The edge population increased from 6.1 to 9.7 plants m^{-2} ($R_0 = N_{t+1}/N_t = 1.6$) and the center population from 5.8 to 22.2 plants m^{-2} ($R_0 = 3.8$). The influence of generation time as discussed earlier has not been taken into consideration here. R_0 for species with multiple year generations is $N_{t+\tau}/N_t$ where τ is the generation time. For this example, the time between life stages is not given and the length of the study is too short to cover a generation. However, these data show how short-term observations on populations do not necessarily match theoretical predictions.

Comparison of Figures 5.10A and 5.12A demonstrates two different patterns of mortality for two different shrubs, *A. suaveolens* in Australia and *C. scoparius* in North America. Once Scotch broom becomes established plants survive well and continue to grow and have increasing fecundity. On the other hand, survival of *A. suaveolens* declined gradually with age and fecundity dropped off after 4 years. Life tables are a convenient way to organize data on plant populations, but it must be kept in mind that they are not static. Conditions influencing survival and fecundity of plants vary all of the time. They are average values and species of plants will vary in their survival and reproduction curves.

Population ecology of vegetatively reproducing plants

Studying the population dynamics of vegetatively reproducing plants is complicated by the difficulty of identifying individuals. For example, *Ranunculus repens* is a common clonal perennial plant with ramets on above-ground stolons (Chapter 4). Stolons grow during the summer and then rot in the autumn leaving independently rooted rosettes (Lovett Doust 1981). The birth rate of the population is determined by the number of ramets produced by rosettes and death rate is the death of rosettes. Lovett Doust (1981) studied the population dynamics of *Ranunculus* in a grassland and a woodland site in Wales by tracking the birth and death rates. She found that the production of ramets per rosette four to six, was similar in both habitats and was independent of density. However, the death rate of rosettes in the summer was density dependent. The death rate of ramets was particularly high in the summer when density was also high. Although *Ranunculus* also produces flowers, seed production in these populations was very low.

Aquatic plants often reproduce vegetatively. Room (1983) studied the population dynamics of the floating water fern, *Salvinia molesta* and developed a model of its population growth. In this case an individual is referred to as a colony composed of ramets connected by a rhizome. He studied *Salvinia* in two conditions, fertile and infertile lakes. In fertile lakes rhizomes grew rapidly and became branched. Rhizomes were brittle and broke apart with repeated branching to form new colonies. In infertile water the rhizomes were tough and unbranched. This allowed nutrients to be reallocated within the colony from old to new ramets. Branching was not related to crowding. The age of plants could be determined by the branching pattern and generations were indicated by the number of branches. The understanding of the ecology of *Salvinia* was important to its biological control, as described in Chapter 7.

The attachment between plant parts in vegetative reproduction can buffer populations against environmental variation. This may be related to their greater occurrence in stressful environments (Chapter 4). Vegetatively reproducing plants can changes their growth pattern, as described for *Ranunculus* in Chapter 4. This may be a more important response to density than is the production of new rosettes or ramets. Comparisons of the population dynamics of sexual and vegetatively reproducing plants is an interesting area for study.

Case study – Diffuse knapweed in British Columbia

Three species of biological control agents were introduced to a population of diffuse knapweed populations in the dry interior region of British Columbia. To determine how knapweed might compensate for the attack of these insects that primarily reduced seed production, Powell (1990a, b) undertook a population study and developed a model to look at determinants of birth and death rate functions. For this he studied knapweed at one location over 3 years. He obtained seedling and rosette mortality rates from twenty-four 0.5 m^2 quadrats of marked and mapped plants. The proportion of mature plants flowering was determined from forty-eight 0.5 m^2 quadrats. The number of seeds per plant was determined from 11, one square metre quadrats.

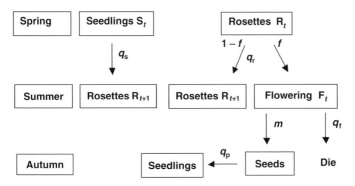

Figure 5.13. Interactions among life stages of diffuse knapweed. q values refer to the mortality occurring between life stages and f is the transition from rosette to flowering. Based on Powell (1990a).

The interactions among life stages of knapweed are shown in Figure 5.13 in which R, F, S, and s are rosettes, flowering plants, seedlings, and seeds respectively and the q values refer to the mortalities of the stages: q_p of seeds, q_s of seedlings, q_r of rosettes, q_f of flowering plants. The transition from rosettes to flowering plants is given by f. The population difference equation for change between years t and $t + 1$ is as follows:

$$R_{t+1} = R_t(1 - f)(1 - q_r) + S_t(1 - q_s) \quad \text{and}$$

$$\text{birth rate} = \frac{S_t(1 - q_s)}{R_t}$$

(recruitment of rosettes)

death rate (total of pre- and post-reproductive

$$\text{mortality}) = (1 - f)q_r + f q_f$$

Each rate variable in these equations, q_r, q_s, and q_f and seedling density, S_t is related to the density of mature plants and, therefore, to determine the equilibrium population density, the point at which the birth and death rate functions intersect, these relationships must be known. The relationship of rosette density to seedling mortality, rosette mortality, proportion of rosettes developing to flowering plants and seedling density are shown in Figure 5.14A–D and the regression equations are given in Table 5.5. The relationship of seeds per flowering plant to rosette density (Figure 5.15) shows a decline with increasing density at densities below 100 plants m^{-2} but great variation without any trend above 200 rosettes m^{-2}. By substituting the regressions of rate variables into the birth and death rate equations Powell determined the birth rate and death rate

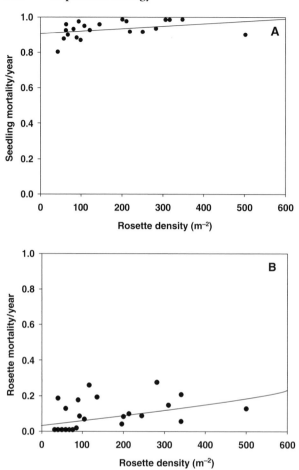

Figure 5.14. Relationships between rosette density and mortality of four life stages of diffuse knapweed measured for a population in White Lake, British Columbia. Based on these the regressions of the rate processes needed to calculate the birth and death rates were determined and are given in Table 5.5. After Powell (1990a). (Figure 5.14 cont. on p. 143.)

functions (Figure 5.16). Both of these relationships are concave with an equilibrium of approximately 77 plants m^{-2}. These relationships show that it will be difficult for a biological control agent to reduce the density of the population. As density declines the birth rate increases as does the proportion of flowering plants that become reproductive. The latter causes the death rate to go up with reduced plant density because almost all flowering plants die after flowering. But the flowering plants will also

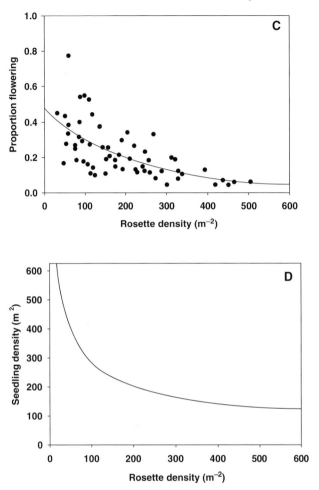

Figure 5.14 (cont.)

have high seed production before dying, and this will result in seedling and rosette production. Thus knapweed populations are well buffered against herbivore attack.

The ability of knapweed to compensate for increased mortality was shown experimentally by Myers *et al.* (1990). In the spring, 30 square metre quadrats were marked out and 0, 40%, or 80% of the rosettes and flowering plants were removed in 10 quadrats. In a quarter of the quadrats the number of seedlings was counted. In the autumn the number of rosettes and flowering plants were counted in each quadrat and the number of seedlings in the quarter quadrat. These experiments were done

Table 5.5. *Summary statistics and regressions for relationships between plant density and f (proportion rosettes flowering), q_r (mortality of rosettes), q_s (mortality of seedlings), $\log_{10}s$ (seed density) for low and high densities (see Figure 5.14) and S_t (seedling density)*

Regression equation	r^2	df	F	Significance
$f = \sin(0.669 - 0.0011R_t)^2$	0.475	1,68	60.7	$P < 0.001$
$q_r = \sin(0.115 - 0.0007R_t)^2$	0.21	1,22	5.79	$P < 0.05$
$q_s = \sin(1.322 - 0.00037R_t)^2$	0.13	1,22	3.19	n.s.
$\log_{10}s = 3.709 - 0.008R_t$ (low)	0.27	1,27	9.85	$P < 0.005$
$\log_{10}s = 1.901 - 0.0031R_t$ (high)	0.004	1,171	0.659	n.s.
$\log_{10}S_t = 3.671 - 0.613\log_{10}R_t$	0.594	1,22	32.21	$P < 0.001$

From Powell (1990a).

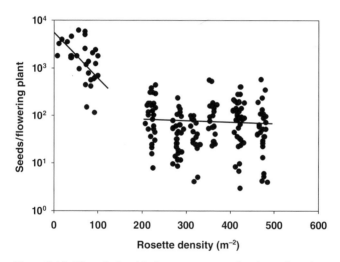

Figure 5.15. The relationship between rosette density and seeds per flowering plant for diffuse knapweed measured for a population in White Lake, British Columbia. The regression for low rosette densities is significant while that for high densities is not (see Table 5.5). After Powell (1990a).

at three sites and results for one site are shown in Figure 5.17. At this site removing mature plants reduced the density but increased the survival of the seedlings.

This case study of diffuse knapweed demonstrates that the ability of plants to respond to density and changes in density. It is a good example of combining experimental manipulation of plant density with

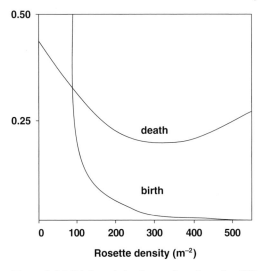

Figure 5.16. Birth and death rate functions for diffuse knapweed based on regressions in Table 5.5 and Figures 5.13 and 5.14. After Powell (1990a).

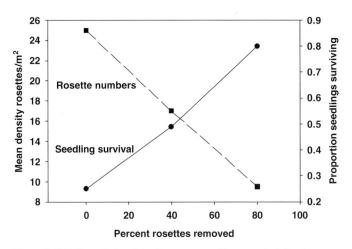

Figure 5.17. The relationships between seedling survival (spring to autumn) and rosette numbers (autumn) of diffuse knapweed resulting from an experimental removal of 0, 40% or 80% of rosettes in the spring. This shows the ability of populations to compensate through increased seedling survival. Data from Myers *et al.* (1990).

quantification of plant growth and survival and population modeling. By measuring the relationships between density and survival, the impacts of different levels of insect attack could be predicted. Experimental removal of plants also resulted in direct observations of the ability of plants to compensate for increased mortality. These types of study of plant population ecology can be useful in determining if biological control is likely to be effective.

Conclusions

Determining what influences the population ecology of plants initially seems very simple. However, it quickly becomes apparent that this is not the case. The impacts of habitat and environmental heterogeneity on populations are complex and plants adapt quickly to different conditions. Recruitment to populations may initially be seed limited or site limited. Self-thinning occurs between the seedling and adult plant stages but the environment determines the total productivity of a site. Limiting factors set average densities and regulating factors are those that respond to plant density. Monitoring populations yields data on changes in population density and in the survival and reproductive parameters that cause the population to change. Life tables provide a convenient format for analyzing population data. Long-term data are important for integrating the environmental impacts on populations, but experiments are also valuable for determining how plants respond to crowding and whether they are seed or microsite limited. Population studies of invasive plant species in their native and introduced habitats can provide information relevant to their ability to spread and to their potential control.

6 · *Introduced plant diseases*

Introduction

With the introduction of plants and plant products comes associated plant diseases. Diseases are particularly insidious because they may show few or no symptoms on the hosts with which they have evolved. However, native species that are susceptible to these new diseases are likely to have very low resistance or tolerance. Introduced plant diseases have probably had stronger impacts on native plant communities than have their original plant hosts. In some cases, introduced diseases have changed the species composition of forests in a major way.

It is often difficult to identify the origins of plant diseases. A survey of potato, rhododendron, citrus, wheat, Douglas fir and kudzu found that an average of 13% of their pathogens were non-indigenous (Schoulties, cited in OTA 1993). Among forest pests that have been introduced on imported live plants are dogwood anthracnose, *Discula destructiva*, *Melampsora* fungus on larch and poplar, and pine pitch canker, *Fusarium subglutinans*, on Monterey pine. Green wood used in packing and green logs and wood chips are potential carriers for plant disease organisms. Imports of raw wood products from distant sources are increasing and so are the risks of introducing new forest diseases (Campbell 2001). One fascinating case is the introduction of the fungus *Cryptococcus neoformans gattii* to coastal British Columbia, Canada. This is a saprophyte of tropical eucalyptus trees that has now established on Douglas fir, cedar and alder trees. This pathogen is capable of infecting mammals, and has been blamed for killing two humans, four porpoises, and 33 other mammals. Although it is often difficult to track the source of a new disease such as *Cryptococcus*, the stories of some introduced plant diseases have been well documented. We describe here five case studies demonstrating the potential impacts of introduced plant diseases, and an overview of biological control using fungi with two case studies. Finally, we consider the role of soil fungi in promoting the invasiveness of plants followed by a

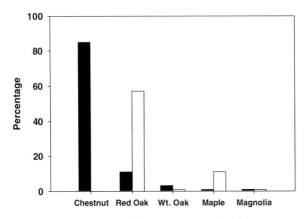

Figure 6.1. Percentage of the canopy (1932) and percentage relative density (1982) of five tree species in a formerly chestnut-dominated forest in southwestern Virginia. Data from Stephenson (1986).

consideration of how the frequency of introduction of plant diseases might be reduced.

Chestnut blight (*Cryphonectria parasitica*)

Chestnut blight, *Cryphonectria parasitica*, is a fungal disease native to Asia. The fungus was probably introduced with nursery stock to New York City in about 1900. Unfortunately, American Chestnut, *Castanea dentata*, a dominant species in hardwood forests in the mid-Atlantic region of the United States and throughout the Appalachian Mountains, was highly susceptible to this disease. The fungus enters the tree through a wound and grows in the cambium, eventually killing the tree within 2 to 10 years by girdling. Roots of the chestnut trees are not influenced by the fungus, and chestnuts have continued to exist as saplings arising from rootstocks. Eventually, however, the saplings are also attacked and die.

From an ecological standpoint the introduction of chestnut blight has been a massive experiment. What happens when a dominant tree species disappears? A number of studies have monitored the changes in eastern hardwood forests following the impact of chestnut blight. In 1932 E. Lucy Braun surveyed a chestnut-dominated forest in southwestern Virginia before blight had reached this far south. Fifty years later Stephenson (1986) resampled these same areas. Chestnuts had disappeared from the forest canopy and red oak, *Quercus rubra*, and red maple, *Acer rubrum*, had increased (Figure 6.1). Surveys of these forests indicate that in addition to the drastic change in species composition following the removal of chestnuts, the forest composition continues to change, partly associated with environmental variation (Abrams *et al.* 1997) and partly with succession (Stephenson and Fortney 1998). Therefore, the response of the forests to the removal of chestnut trees may still be in progress. Oaks and maples have primarily filled the open niche created by the loss of the American chestnut.

Some large chestnut trees have been found that are resistant to the blight fungus (Brewer 1995), and efforts continue to develop resistant strains of chestnut trees through cross breeding of American and Chinese chestnuts. In addition, endogenous bark fungi from healthy chestnut trees are being explored as possible biological control organisms (Tatter *et al.* 1996). It seems likely, however, that the American chestnut will not return to dominance in the hardwood forests of eastern North America for centuries, if ever. A further interesting interaction may occur in these areas in the near future. Attacks by the introduced gypsy moth, *Lymantria dispar*, are increasing. Oaks are the preferred host species of the moth larvae, and maples may again gain an advantage over oaks in the succession process. This is a clear example of the continuing dynamic change in succession discussed in Chapter 3.

Joint introductions – common barberry and wheat stem rust

Common barberry, *Berberis vulgaris*, was brought to North America from Europe by early settlers. Barberry had several characteristics that may have led to its introduction. First, it had been widely used in Europe as a medicinal plant. In addition, the wood of barberry was long, straight and light and made excellent handles for tools. The dense growth and sharp spines of barberry made it an excellent hedge for corraling livestock. The fruit of barberry was valued for sauces and preserves and finally the bark provided a source of a yellow dye. Not surprisingly, barberry and wheat became close associates as agriculture spread in North America in the early 1800s. Plants were spread by farmers, and seeds were dispersed by birds, mammals and water.

Even before it was introduced, barberry had been recognized in Europe as an alternate host for the stem rust fungus. The earliest laws against growing barberry in the Americas were passed in 1726 in New England, but there seems to have been little enforcement. In 1916 a disastrous epidemic of stem rust on wheat in the mid-west led to the development and enforcement of laws against growing barberry in wheat growing states, and barberry eradication programs were established (Roelfs 1982). Between 1918 and 1975 over 100 million barberry plants were destroyed, but some bushes still survived in isolated, woodland sites distant from cultivated land. New bushes also arose from seeds in the seed bank and a current resurgence of barberry is now in progress. An account of the eradication attempts at one early farmstead in Minnesota describes the removal of all plants arising from a barberry hedge in the 1920s, the 1940s and from then until 1975 at 7-year intervals. But in 1995 three hundred bushes up to 15 feet tall occurred on this farm. Suppression was possible, but not eradication.

Stem rust epidemics occurred in 1937 and in 1953 and 1954. Both the depletion of barberry and the development of strains of wheat resistant to stem rust have reduced the problem from this fungal disease. However, with the resurgence of barberry new epidemics can be expected. The persistence of fungus on alternative hosts allows the continual development of new races and these can rapidly overwhelm the resistance of new wheat varieties.

This story has many fascinating dimensions. First, barberry was brought by farmers to North America even though it was known, at least by some, to be a host for an agricultural disease. This is not surprising given that there was probably no restriction on plant importations in the 1700s. Secondly, the eradication campaign was successful at reducing barberry density and distribution. The decline in wheat stem rust was most apparent in the initial stages of barberry removal when plants around the wheat fields were removed (Figure 6.2). However, this is an example of how introduced species can be suppressed through considerable diligence, but eradication cannot be complete. Some plants and seed banks remain to initiate re-establishment. It is typical that we become complacent when a problem is suppressed and one of the necessities of successful eradication is to have the will and the resources to get the last individual (Myers *et al.* 2000). In the long run the solution for the wheat rust problem may be found in genetic modification of wheat. However, it is also important to consider the role of alternate hosts for maintaining disease organisms in agricultural systems. The existence of an alternate host allows new biotypes of fungus to develop and these provide genetically variable strains of rust with the potential to overcome the resistance of selected wheat varieties.

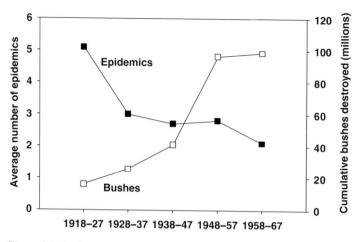

Figure 6.2. As the cumulative number of destroyed barberry plants increased, the number of epidemics of wheat stem rust declined. Impacts were strongest in the early part of the eradication campaign. Data from Roelfs (1982).

Sudden oak death and rhododendrons

Since 1995 the deaths of three native oak species in California have reached epidemic levels. The impact of the disease on the native oaks led to its being called sudden oak death or oak mortality syndrome. Once symptoms occur, the tree dies within six to eight weeks. Five years were required before the organism associated with sudden oak death was identified. During this time there were suspicions that the cause might be the same fungus responsible for chestnut blight. However, this turned out not to be the case. By 1999 the numbers of dead oaks in California had continued to climb to the tens of thousands (Garbelotto *et al.* 2001). In 2000 a new species of the fungus *Phytophthora* sp. (technically a brown alga, but still called a fungus) was identified. Shortly after, following a lead from colleagues in Europe, the same fungus was identified on rhododendrons in a nursery. What led David Rizzo and Matteo Garbelotto to look at rhododendrons was reports of a similar pathogen on rhododendrons in Germany and The Netherlands. Oaks are not attacked in these countries, however (Kan-Rice 2001).

While there are many details to work out about this new disease, its association with rhododendrons as a carrier is crucial. Quarantine regulations are now in place to prevent the movement of rhododendrons unless they have been certified to be disease free. Considerable effort is now directed to tracking the movement of the disease with GIS mapping (see Appendix). Monitoring the spread is facilitated by the obviousness of dead oak trees. If this disease continues to spread it is likely that the prevalence of oaks in dry areas of the western United States and southwestern Canada will be severely threatened. This is another example of the serious impact of a plant disease being imported to a new continent (Gerlach 2001).

White pine blister rust, *Cronartium ribicola*

White pine blister rust is a disease of five-needled pines caused by the fungus, *Cronartium ribicola*. The life cycle of the rust alternates between five-needled white pines (genus *Pinus*) and shrubs of the genus *Ribes*. The history and biology of white pine blister rust have been reviewed by McDonald and Hoff (2001).

In 1854 white pine blister rust was first recognized in the Baltic states on black currant, *Ribes nigrum*, red currant, *R. rubrum*, gooseberry, *R. palmatum*, and on planted forests of North American eastern white pine, *Pinus strobus*. Rust later appeared in planted pine forests in Finland, East Germany and Denmark, and by 1900 it had spread through northern Europe. White pine blister rust infections were most severe in areas of forests planted with eastern white pine, a species that had been introduced to Europe in the 1700s. The source of the rust was tracked to the introduction of Siberian stone pine, *Pinus cembra*, to northern Europe. The introduction of the exotic rust led to the elimination of the exotic host, eastern white pine.

The interesting twist to this story comes because the North American, eastern white pines could no longer be grown in Europe. Nurseries were producing seedlings and when European markets closed, they shipped seedlings back to North America. Foresters were warned to check the seedlings for signs of rust, but they mistakenly looked for rust infection on the stems and not on the needles where in fact the infection begins. From 1890 to 1914 millions of seedlings were sold to North American markets. In 1906 blister rust was found on *Ribes* in New York and in 1909 infection of pines initiated from European seedlings, now widely spread in eastern North America, was confirmed. Eradication of the secondary host, *Ribes*, in the vicinity of white pine forests was attempted but found to be impossible.

The spread of white pine blister rust to western North America was first recognized on black currants on Vancouver Island and in the lower

Figure 6.3. Distribution of white pine blister rust on whitebark pine, *Pinus albicaulis*, in western North America. After McDonald and Hoff (2001).

Fraser Valley in British Columbia in 1922. Eventually the original source was traced to an importation of 1000 seedlings from France to Vancouver in 1910. By 1922, of 180 trees remaining from this importation, 68 were infected with rust. The spread of rust appears to be associated with 'wave' years in which cool moist spring weather is followed by frequent cool, moist infection periods during the summer. The rapid spread of the rust is shown in Figure 6.3. Mielke (1943) concluded that three years after the introduction of the infected white pines to Vancouver the blister rust spread to native *Ribes* 380 km away.

Surveys of whitebark pine in southwestern Canada and western United States show levels of mortality from rust in some sites to be up to 100% (Kendall and Keane 2001). Some resistant white pine trees have been found and these have been used to grow seedlings for restoration. However, in northwestern North America rust is one of several factors causing serious declines of whitebark pine, *Pinus albicaulis* (Tomback and Kendall 2001). Whitebark pine is considered to be a keystone species in high elevation sites where after establishment it modifies the harsh environment for other species. Its large and nutritious seeds are an important food source for birds, particularly the Clark's nutcracker, *Nucifraga columbiana*, as well as for squirrels, other small mammals and even large mammals such as bears. In addition Clark's nutcrackers are important as seed dispersers for other pines. With declines of nutcrackers and policies to suppress forest fires, in addition to white pine blister rust infection and attacks by mountain pine beetles, *Dendroctonus ponderosae*, the white pine ecosystem may become so seriously damaged that restoration will be impossible.

This is another example of the impact of an introduced plant disease on a forest ecosystem. It differs from the chestnut blight but is similar to the wheat stem rust and oak sudden death situation, in that a widespread secondary host is a major component of the maintenance and spread of the disease. The ability of rust fungi to spread great distances in the wind makes it impossible to stop the disease once it has been introduced to a new area. In addition to selecting for resistant trees to be used in restoration, another area of research involves the identification of endophytes (cryptic fungi growing inside plant tissue) on needles (Bérubé and Carisse 2000). Endophytes have been found that inhibit *C. ribicola* on both pines and *Ribes*. Developing these organisms into useful control agents is still a major undertaking but holds some promise.

Pandemics of Dutch elm disease, *Ophiostoma ulmi* and *O. novo-ulmi*

The history of the demise of elm trees in the northern hemisphere has recently been summarized by Brasier (2001). Elm trees in the genus *Ulmus* occur widely throughout temperate regions of the Northern Hemisphere. Their major enemy is a wilt disease caused by ascomycete fungi of the genus *Ophiostoma*. Two major pandemics of what has become known as Dutch elm disease, named because much of the early work was done in The Netherlands, have had major impacts on elms throughout their range. The fungus is spread by the activities of elm bark beetles in the genus *Scolytus*. Thus the potential for spread is determined by the movement of infected wood and by the flight of contaminated beetles.

The actual site of origin of the fungus is unknown, but it is thought to have come from Asia. The first pandemic of Dutch elm disease started in northwestern Europe around 1910 and it spread east from there into southwestern Asia. The fungus was introduced into Britain and North America in 1927. This first wave of disease was caused by *O. ulmi*, and the initial outbreak declined in the 1940s after killing 10–40% of the elm trees.

In the early 1970s a second outbreak of a more severe form of Dutch elm disease began in England and spread to neighboring Europe. It was discovered that this was associated with a new, more virulent pathogen *O. novo-ulmi*. Later research showed that this form of the fungus had

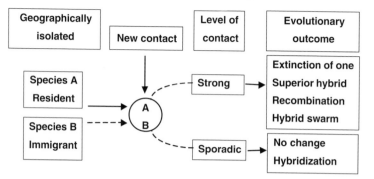

Figure 6.4. The potential for evolutionary change arises when two or more strains of disease organisms come together in a new zone of contact. The evolutionary outcomes will be influenced by the level of contact of the different strains and the degree of exchange of genetic information. Modified from Braiser (2001).

actually been around since the 1940s and it occurred in a number of widely spread locations. In fact, its introduction to Britain was from Canada and arose from a shipment of diseased elm logs in the early 1960s. The spread of these two strains of pathogen has been directly linked to the movement of infested elm timber within and between continents. The more virulent *O. novo-ulmi* has caused catastrophic mortality of mature elms, over 30 million trees having been killed in the United Kingdom and hundreds of millions of elms having died in North America. When the distributions of the two pathogens overlap *O. novo-ulmi* replaces *O. ulmi*.

Braiser (2001) studied the interactions between the two strains of fungus and their potential for hybridization, recombination and evolution. He has outlined some of the potential evolutionary outcomes of new interactions between two formerly isolated pathogen species (Figure 6.4). Contact between formerly isolated strains of diseases can create the opportunity for the development of new hybrids and the potential for evolution to new superior strains. Braiser points out that molecular protocols will be needed to identify hybrid disease organisms and their movements. Disasters are waiting to happen with the movement of plant diseases. The potential for variation among strains and for rapid evolutionary change of pathogens should be recognized in quarantine regulations.

Introduction of fungi for biological control of weeds

If fungal diseases can have such major impacts on susceptible hosts it would seem they would have great potential to be developed as biological controls. A major difference between the introduction of fungi to new locations and the use of fungi for biological control is that in the former new, susceptible hosts are infected and in the latter disease–host interactions are being re-established. Newly challenged hosts are likely to be more susceptible and thus suffer greater consequences than are formerly associated hosts (Hokkanen and Pimentel 1984, Chapter 7). One approach for using fungal disease in weed control is to develop a mycoherbicide based on a fungus that is obtained from the plant, cultured and sprayed onto the plant as an inundative biological control agent (Templeton and Trujillo 1981, Charudattan 1991, Greaves 1996, Greaves *et al.* 1998, Cousens and Croft 2000). There, has been considerable effort in this area of research, but thus far long-term successes, at least in the economic sense, seem rare (Morin 1996). McFadyen (2000) gives three examples in which indigenous fungi have been used successfully as mycoherbicides: control of northern jointvetch, *Aeschynomene virginica*, in the USA by *Colletotrichum gloeosporoides*; of milkweed vine, *Morrenia odorata*, with the fungus *Phytophthora palmivora* in the USA and of broomrape, *Orobanche ramosea*, in the Ukraine and Hungary using *Fusarium* spp. Recently the potential control of kudzu, *Pueraria lobata* (Box 2.5) with a mycoherbicide has been reported (Boyette *et al.* 2002). Kudzu, a leguminous vine from Asia that was spread widely in southeastern USA in the early 1900s now blankets large areas and is a major pest. The fungus *Myrothecium verrucara*, isolated from sicklepod, *Senna obtusifolia*, has been shown to have excellent biocontrol potential for several weeds including kudzu if applied with a surfactant that increases wetting of foliage and penetration of the fungus into stomata. Sprays with fungus alone did not kill kudzu plants.

There is considerable interest in developing fungi as mycoherbicides and several products have received patents (Evans 2002). In more recent studies, fungal pathogens have been used to reduce the competitiveness of agricultural weeds as well as to initiate disease epidemics, by combining the fungus with fungal phytotoxins (Müller-Schärer *et al.* 1999).

Fungi also have potential as classical biological control agents on introduced weeds. Although they have often been rejected for their lack of host specificity, Evans (2002) lists 23 programs that have involved the release of fungi. Of these, eight have been at least partly successful and some very successful at reducing the density of the target host plant (Table 6.1 and Chapter 7) and two of these programs are described in more detail below. Recent programs have not been properly evaluated, so this success rate could improve. One of the most dramatic successes is the control of Noogora burr, *Xanthium strumarium*, with an accidentally introduced fungus, *Puccinia xanthii* (see Chapter 7). Whether this 'accidental

Table 6.1. *Some of the 23 biological control of weed programmes involving the introduction of fungi as listed by Julien and Griffiths (1998) and Evans (2002)*

Host	Native	Introduced	Fungus	Impact
Ageratina riparia (mistflower)	Mexico	Hawaii	*Entyloma ageratinae*	Defoliation substantial control
Carduus tenuiflorus	Europe	South Africa USA	*Puccinia carduorum*	Reduced seeds Established
Chondrilla juncea (skeleton weed)	Europe	Australia USA Argentina	*Puccinia chondrillina*	Controls one strain of weed
Galega officinalis (goat's rue)	Europe	Chile	*Uromyces galegae*	Established Spreading
Acacia saligna	Australia	South Africa	*Uromycladium tepperianum*	Established Spreading Effective
Rubus constrictus *Rubus ulmifolius*	Asia, Europe	Chile	*Phragmidium violaceum*	Weed density decreasing
Ambrosia artemisifolia (common ragweed)	North America	Russia	*Zygogramma suturalis*	Unknown
Cryptostegia grandiflora	Madagascar	Australia	*Maravalia cryptostegiae*	Established Spreading
Xanthium strumarium	USA	Australia	*Puccinia xanthii*	Accidental introduction Spectacular control
Rubus fruticosus	Europe	Australia	*Phragmidium violaceum*	Causing decline
Clidemia hirta	Panama	Hawaii	*Colletotrichum gloeosporioides*	Dieback reported

introduction' was really farmers taking things into their own hands in a program they saw as proceeding too slowly will never be known. Another accidental introduction was that of *Phragmidium violaceum* from Europe onto blackberry, *Rubus fruticosus*, in Australia. Again, this agent is reported to have accelerated the decline of the host population. These accidental escapes of fungi show that, although quarantine procedures are in place, unexplained escapes can still occur.

We next discuss two cases of successful biological control of introduced weeds with fungi.

Uromycladium tepperianum on *Acacia saligna* in South Africa

Acacia saligna is native to Australia but has become a troublesome weed in the Cape fynbos floristic region of South Africa where it forms dense stands in conservation and agricultural areas. Here it replaces native vegetation and interferes with agriculture. Owing to its host specificity, a gall-forming rust fungus, *Uromycladium tepperianum*, was introduced to South Africa from Australia in 1987 and became established at a number of locations throughout the range of the weed. A monitoring program was established in 1991 to evaluate the impact of the fungus (Morris 1997). In 1991 the number of galls per tree was relatively low but infection increased rapidly in subsequent years. During the course of the study between 1991 and 1996 the density of trees at all sites declined to less than 20% of that before the fungus was introduced (Figure 6.5). In association with this the number of seeds in the seed bank declined. Thus within 6 to 7 years of the establishment of the fungus the density of the weed was reduced by 80%. Fires are a regular occurrence in many of the infested sites but some trees appeared to persist following fires and maintain the fungus. Post-fire seedlings became infected. It is interesting that in this case of successful biological control the host is a tree. Unlike the cases described above, no secondary hosts are involved. Spread of the disease was rapid.

Puccinia chondrillina on *Chondrilla juncea* in Australia

Skeleton weed, *Chondrilla juncea*, a native of the Mediterranean region of Europe, became an important agricultural weed in southeastern Australia. Severe infestations can reduce cereal grain yields and make harvest difficult. Three distinct forms of the species occur in Australia, and these can be distinguished primarily by the shape of their rosette leaves. In the 1960s the thin-leafed form was the most widespread and the two other forms had limited distribution. In 1971 a rust fungus from the native habitat

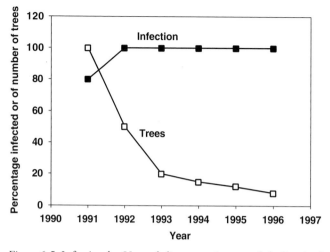

Figure 6.5. Infection by *Uromycladium tepperianum* and decline in the percentage of *Acacia saligna* trees remaining in one of the sites (Coppul) monitored by Morris (1997) following the introduction of fungus in 1989. Data from Morris (1997).

of skeleton weed was introduced to New South Wales, and rapidly established and spread (Cullen *et al.* 1973). The common, narrow-leafed form of the weed was very susceptible to the rust while the other two forms were more resistant. The impacts of the rust were monitored in roadside populations and in most locations between 1968 and 1980 the susceptible form decreased while the more resistant forms increased in frequency (Burdon *et al.* 1981) (Table 6.2). Experiments showed that the rust changed the competitive interactions between the different forms of skeleton weed.

It is unclear if this example is a case of successful biological control using a fungus or a good example of how resistance to disease can prevent successful control. More recent surveys of host and pathogen diversity of *C. juncea* and *P. chondrillina* in their native range in Turkey found eight resistant phenotypes among 19 host lines and seven distinct pathogen types among 15 isolates. There was no obvious relationship between the degree of geographic separation of pathotypes and their virulence on *C. juncea* lines (Espiau *et al.* 1998). Because host specificity is a major concern in biological control introductions, care is taken to introduce only a single strain of a fungus. The lack of variation means that plants resistant to that strain will have a selective advantage and, although initially promising, control can be temporary. *P. chondrillina* also has been introduced on skeleton weed in California and is considered to have been highly

Table 6.2. *Percentages of skeleton weed genotypes identified by their leaf shapes (narrow, intermediate and broad), in roadside populations in NSW, Australia, before and after the release of the rust,* Puccinia chondrillina. *The narrow-leaf form is most susceptible to the fungus*

	1968			1980		
Population	Narrow	Intermediate	Broad	Narrow	Intermediate	Broad
1	100	0	0	88	0	12
2	100	0	0	99	1	0
3	98	0	2	5	90	5
4	67	30	3	3	78	19
5	62	24	14	3	80	17
6	51	10	39	8	10	82
7	21	1	78	11	13	76

Data from Burdon *et al.* (1981).

effective. From there the fungus has apparently spread accidentally to Canada where its impact has not been evaluated (Evans 2002).

The potential role of soil microbes in invasiveness

Soil microbes will vary between native and exotic habitats and the potential role of these on the invasiveness of species has recently been investigated (Klironomos 2002). To determine if there was a positive or negative feedback on plant growth from the soil in which plants grew, Klironomos grew plants from seeds in soil that had previously supported plants of that species, or previously supported other plant species. Five rare plant species from old-field (meadow) sites in Ontario and Quebec, Canada, had reduced growth when grown in their own soil while four of five introduced species had increased growth when grown in their own soil (Figure 6.6). To test if these reactions were caused by soil pathogens and for the impacts of mycorrhizal fungi, soil extracts were taken to separate these two components and were tested on the same plant species. Native species were strongly influenced by pathogens from soil with a history of the same plant species but introduced species were not. Both native and introduced species generally responded in a positive manner to mycorrhizal fungi. Studies were not done with the invasive species in their areas of origin and this should be the next step in the investigation. However, these experiments suggest that the natural communities of soil pathogens may be less virulent against non-indigenous species and,

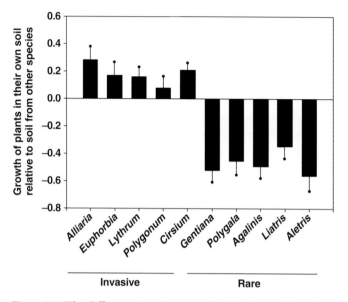

Figure 6.6. The differences in plant growth of native and introduced plant species grown in soil that previously supported the species and soil that had supported other species. Bars represent means ± SE. After Klironomos (2002).

therefore, that a lack of soil pathogens might contribute to the invasiveness of some introduced plants.

The comparisons in this study were between rare native species and common introduced species. Clearly more work should be done to test the reactions of a wider range of native species, including common ones, to soil pathogens. However, the results so far suggest that soil pathogens might have potential to control introduced weeds and might also cause introduced plant species to become invasive. The host specificity of soil pathogens would be a determining factor in whether their development for weed control could ever be seriously considered.

Preventing the introductions of plant diseases

The impacts of foreign plant diseases can be enormous, and yet screening for pathogenic 'hitchhikers' is extremely difficult. Not only live plants but also green wood and soil can be the sources of the next pathogen capable of eliminating a forest species or destroying an agricultural crop. Campbell (2001) points out that a 'science based' phytosanitary program should reflect the seriousness of the threat of introduced plant diseases. Biological invasions are essentially irreversible. However, current regulations under the World Trade Organization prevent nations from using phytosanitary safeguards to protect domestic agriculture from foreign competition and thus, an Agreement on the Application of Sanitary and Phytosanitary Measures has been adopted. This dictates that restrictions on importations should not be greater than necessary. What is the appropriate level of risk that should be accepted? The acceptable level of risk for a country must be supported by a specific risk assessment, and yet predicting what fungus might be on the next shipment of logs, and what impact it may have is extremely difficult. Currently, in most countries trade takes precedence over quarantine, and we can anticipate continuing examples of species being driven almost extinct by introduced diseases.

The recent US National Academy of Sciences report on predicting the invasions of plants and plant pests (NAS 2002a) lists characteristics of pathogens that might indicate the likelihood of their arrival, establishment and invasiveness (Table 6.3). As with attempts to predict the invasiveness of plants, this report suggests risk assessment as a management strategy. In addition to the characteristics listed in Table 6.3, the consequences of introduction would also be considered, such as economic and environmental impacts. They site as an example the qualitative pest risk-assessment of importation of solid wood packing based on a 3 × 3 matrix of factors: the likelihood of introduction being high, medium or low and the consequences of introduction being high, medium or low. The potential for entry is high if there is a history of repeated interceptions, or if two of the three following characteristics apply: (1) the pathogen is likely to survive transport, (2) it is not likely to be separated from its host, and (3) it is difficult to detect. Obtaining information for risk assessment of pathogens that are known is difficult but when one is dealing with unknown organisms it becomes nearly impossible. A more realistic solution would be to prohibit the importation of any wood packing unless it has been treated.

Conclusions

The diversity of diseases that attack plants is enormous. For example, in an evaluation of potential fungal and bacterial agents of Canada thistle, *Cirsium arvense*, growing in the Canadian prairies, 287 pathogenic fungi were isolated (Bailey *et al.* 2000). Most of these had little impact on their host plant. However, the environmental and agricultural impacts of some introduced diseases make it important to consider regulations on the movement of plant material.

Table 6.3. *Characteristics of pathogens that might indicate the likelihood of their arrival, establishment and invasiveness*

Characteristic	Information needed
Arrival	
History	Evidence of past introductions?
Rate of movement	1 Commonly associated with imported products?
	2 Volume of importation high?
	3 Wide geographic range?
	4 Can be moved passively?
Survival	Can the pathogen survive transit?
Escape from safeguards	Can the pathogen be recognized and detected?
Establishment	
Environmental suitability	Are habitats at origin and destination similar?
Host finding	1 Susceptible hosts present?
	2 Hosts with spatial and temporal synchrony?
	3 Is a vector present if necessary?
	4 Is the pathogen very virulent?
Overcoming environmental and demographic stochasity	1 Can it survive adverse condition?
	2 Is the inoculum pressure high?
Invasion	
History	Detrimental effects on plants elsewhere?
Host-habitat availability	1 Are susceptible hosts widespread?
	2 Have a diversity of pathogenic traits been demonstrated?
Dispersal	Can it spread rapidly and widely?
Growth	Does it have a high reproductive capacity?

Adapted from NAS (2002a).

The development of particular diseases for biological control will require extensive screening for both host specificity and for impact. New molecular techniques are valuable tools for identifying and screening plant pathogens (Becker *et al.* 2000) and for tracing the spread of introduced plant diseases. There is currently considerable interest in the potential impact of plant diseases for biological control to be used as mycoherbicides, but progress so far has been slow. What makes the difference between the pathogens that can devastate plants over a wide geographical range and those that have almost no impact? Answering this question is a challenge for future plant biologists. Finally, it is likely that microorganisms are going to be found to have crucial roles in plant-to-plant interactions. These may be very important to determining the invasiveness of plant species.

7 · Biological control of introduced plants

Introduction

Densities of invasive, introduced plants are generally higher in the exotic habitat than in their native habitats. Possible reasons for this have been discussed in Chapter 3 and include a lack of specialist herbivores and new competitive interactions with other plant species in the exotic habitat. If the success of invasive weeds is due to a lack of specialized herbivores and diseases, the introduction of natural enemies from the native habitat should redress the problem. This rationale, however, produces an ethical dilemma. Should more foreign species be introduced to adjust the balance between native and introduced plant species?

A review of biological control in Canada showed that on average five to seven species of natural enemies were introduced for every exotic weed for which biological control was attempted. Of these, only 10% had any impact on host density. This ratio of introduced agents to targets is approximately 2.5 to 1 in other studies (McFadyen 2000). However, some biological control programs have involved a very large number of introductions of natural enemies. For example, over 20 species of natural enemies have been introduced in largely unsuccessful attempts to control *Lantana* (Broughton 2000) and 50 species of natural enemies were introduced early in the biological control programme against *Opuntia* cactus in Australia (Mann 1970). The practice of biological control increases the number of introduced species, and in this way has the potential to increase the ratio of exotic to native species. The impact of these species can be even greater if they attack and reduce the density or distribution of native plant species (Louda *et al.* 1997). On the other hand, the high density of an established invasive weed can suppress native vegetation and reduce at least local biodiversity.

The Neo-European countries (see Chapter 2) have been the most involved in biological control programs against weeds, being led by the United States with projects on 54 target weeds, Australia 45, South Africa

28, Canada 18 and New Zealand 15 (McFadyen 1998). Other countries with classical biological control programs are Malaysia, Thailand, India, Indonesia, Vietnam, Papua New Guinea, China, Uganda, Zambia, Tanzania, Kenya, Ghana, Côte d' Ivoire, Benin, Argentina, and Chile. In Europe, only the former USSR has made deliberate releases for biological control.

Carrying out a biological control project is usually a long process, often lasting over 10 years. Because biological control involves the introduction of species, government regulations must be met and, since control agents can move across borders, international agreements are necessary. Therefore, it is a procedure that cannot be undertaken lightly, and over the years considerable effort has gone into making biological control of weeds as efficient and safe as possible. There are costs and benefits to the practice and the need for biological control increases as more plants are introduced to new areas.

In this chapter we consider the following important questions about biological control of weeds: How successful is biological control? Can we predict what characterizes successful biological control agents and suitable species for control? How many agents are necessary for successful biological control? How safe is biological control and what is the potential for non-target impacts of biological control?

How successful is biological control?

Biological control of introduced weeds has been practiced since early in the twentieth century. The details of carrying out a biological control of weeds program are outlined by Hartley and Forno and the steps are summarized in Box 7.1. The first, and one of the most successful biological control programs, was the control of prickly pear cactus in Australia by the *Cactoblastis* moth (Box 7.2). In this case success was measured as the reduction of the density of the target plant and the restoration of land to productive agriculture. However, success can be evaluated at different levels. Anderson *et al.* (2000) define seven different types of success that can be achieved with biological control.

Biological success is the reduction of the density of the target species to the point that it is manageable in the landscape. As will be discussed further below, biological success requires quantification of the density of the target plant before and after control, and an understanding of management procedures.

Box 7.1 · *Steps in a biological control of weeds program based on Hartley and Forno (1992).*

1 Selecting a target weed – Determination of whether a biological control program should be initiated starts with investigations in the country of introduction. Factors to be considered are chances of success, cost effectiveness and conflicts of interest. Often a vocal group such as a cattlemen's association or even a very forceful individual will lobby government agencies to initiate a biological control program.

2 Initiation – A program begins with a review of the literature and compilation of data on taxonomy, biology, ecology, economics, distribution, and known natural enemies of the target weed. It is very important to identify the target weed correctly and to know its area of origin.

3 Foreign exploration – This involves the search for potential agents in the native range, documentation of the insects and pathogens that attack the weed and consideration of the biological character-istics of these natural enemies. Although this sounds exciting, for-eign explorations are often carried out under difficult conditions. Collected insects must be kept alive and sites are often remote.

4 Ecology of weed and natural enemies – Studies of the target weed, its relatives and natural enemies in the native habitat provide valu-able information and are necessary for the documentation required to obtain an import permit. Experimentation by creating high densities in the native habitat could be used to identify potential agents. This potentially valuable tool has not been used often.

5 Host-specificity testing – To prevent non-target impacts, relatives of the target plant species and major crops in the area of introduc-tion are exposed to the insect or pathogen to determine if they are suitable for oviposition by the insects, attack, or infection and if the insects can complete development on them. This is done in the native habitat or under quarantine. Evaluations are done with only one plant species being presented to the insect with no choice, and with choice experiments in which several species are pre-sented at once. A brief review of procedures is given by Kluge (2000).

6 Preparation of documentation for import permits – Regulatory agencies consider requests for importation of biological control agents. International cooperation is required if agents are capable

of moving across borders. This process can take considerable time (years).

7 Importation and quarantine – Agents are usually reared for one generation in quarantine to check for accompanying parasitoids or diseases.

8 Mass rearing and release – To obtain sufficient numbers for release, agents are frequently reared either in the laboratory or in field cages. The success of introductions is often related to the number of individuals introduced. However, it is important to avoid adaptation to the laboratory environment.

9 Evaluation and monitoring – Field studies on the establishment and spread of species are necessary to evaluate the impact of biological control agents. Control sites should be established and impacts evaluated in terms of the density of the target weed. This is often not done because it is hoped that the impact will be so dramatic that quantification will not be required. It is also difficult to get funds for this stage.

10 Distribution – If agents attain high densities in the initial release sites they can be collected and redistributed to new sites. This may be done with 'field days' in which farmers are invited to come and collect agents to redistribute on their own land. Records should be kept of introduction sites and this is now facilitated with GPS (see the Appendix) that allows coordinates to be obtained for geographic areas.

Historically, biological control programs were carried out by government agencies but with increasing privatization, consulting companies are now often involved.

Box 7.2 · *Prickly pear cactus and the* Cactoblastis *moth.*

Prickly pear cacti are native to tropical America. The first prickly pear to be introduced to Australia was *Opuntia monacantha*, imported from Brazil by Captain Arthur Phillip in 1788. It is thought that this was done with a view to developing a source of red dye from the cochineal bugs that feed on the cactus, to provide a source for the production of red coats for the British Army. The growing of potted cacti was popular at this time and thus plants were collected from the vicinity of ports and carried around the world. Two species of *Opuntia* that were introduced, but for which taxonomy is still uncertain, are *O. stricta*, from Chile, and *O. inermis*, from USA. The first *O. inermis* was brought from Galveston, Texas, to Australia in 1839 as an ornamental plant. *O. stricta* and *O. inermis* were subsequently widely spread in Australia as ornamentals and hedgerow plants. They were also valued as a source of fruit and as forage for cattle during drought. By 1916 these prickly pears had invaded 60 000 000 acres and were estimated to be spreading at a rate of 1 000 000 acres per year (Mann 1970). Homesteads had to be abandoned when pastures and fields were overrun by cactus that could reach heights of 1–2 m. This environmental disaster lead to the search for insects, native to the Americas, that could be imported as biological control agents.

The first explorations for potential biological agents were initiated in 1912 by the Queensland Prickly Pear Travelling Commission. These introductions were not successful and all efforts were put on hold until the end of the First World War. In 1924 exploration began again and Alan Dodd collected larvae of the moth *Cactoblastis cactorum* in Argentina. *Cactoblastis* moths (Figure 7.1) attack a variety of species of *Opuntia* cacti and occur over a wide range in Argentina, Paraguay, Uruguay, and Brazil. They lay their eggs in spine-like 'egg sticks' of 60–100 eggs piled one on top of the other. The larvae hatch and feed gregariously in the leaf or cladode of the cactus. The damage associated with their feeding allows fungal disease to develop, and heavily attacked plants die. Dodd collected eggs from 3000 moths and these were packaged for a 10-week journey by sea to Australia. Between 1925 and 1929 moths were reared and over 390 million eggs were distributed in eastern Australia. Field populations were so successful that at one center, 300 million eggs were collected for redistribution in February 1930. Within 3 years of the introduction of the *Cactoblastis* moth, the decline of cactus was well under way (Mann 1970). The success of *C. cactorum* was recognized by grateful Australians with the building of a memorial hall in Boonarga, Queensland (Figure 7.2).

Cactoblastis moths were subsequently introduced to other countries as a biological control agent for cactus. Release sites included the island of Nevis in the Caribbean, where the target was native *Opuntia* species that were pasture weeds (Simmons and Bennett 1966). This event began the transition of the reputation of *Cactoblastis* from biological control hero to biological control demon. In 1989 *Cactoblastis* were found in Florida attacking native *Opuntia* species. One of the six native

Figure 7.1. Cactoblastis cactorum introduced for the control of *Opuntia* cactus in Australia.

species of *Opuntia* attacked by *Cactoblastis* in Florida, *O. corallicola*, existed as only 12 individuals in the Florida Keys. Thus the threat of moth attack on this species' survival is considerable. Subsequently *C. cactorum* has been found in Georgia and has demonstrated a rate of spread similar to that observed in Australia after its introduction (Stiling and Moon 2001).

It is still uncertain if moths flew to Florida from the Caribbean or if they were brought in with the ornamental cactus trade (Pemberton 1995). Will the moth continue to spread across the Gulf states and into Mexico, the center of cactus diversity? In Mexico *Opuntia* cactuses are used as food and for fodder. They are so important to the Mexican way of life that they are featured on the Mexican flag. It seems that the causing outbreaks. The story of prickly pear cactus and its control is one of intentional introductions of plants and insects causing major and unexpected impacts. It is also a story of how enthusiasm for biological

Figure 7.2. John Monro standing in front of the Cactoblastis memorial hall in Boonarga, Queensland.

success can override attention to ecological relationships. Targeting native species with an introduced biological control agent can open a Pandora's Box that can never be closed. Thus introductions must be made with caution.

Ecological success is the stopping or reversing of the progression of an invading plant through the use of biological control agents and other integrated pest management tools. Ecological success should have no direct detrimental effects on the ecosystem, or at least any negative effects should be outweighed by their overall benefits. Ecological success takes into consideration that there can be non-target or indirect impacts of control agents (Cory and Myers 2000). Even the reduction in weed plant density can have other impacts on the environment such as increasing erosion, reducing bees for pollination, allowing the invasion of other species, or removing plants that are used for food or nesting by native animals (Zavaleta *et al.* 2001). From a rancher's perspective, removing the species is sufficient, but land managers have the mandate to improve the productivity, quality, and uses of the land. Therefore, ecological relationships are an important dimension of successful control.

Scientific success is achieved with the acquisition of knowledge through scientific investigation to gain understanding of biological control and the effects of agents on the host plants and their density in the context of the ecosystem. Even if control is not achieved, an improved ability to predict what will and what will not work in biological control is a measure of scientific success. Research of this kind might consider the additive or synergistic effects of insects and pathogens and should contribute to the understanding of how weed control programs can be enhanced. Scientific success can be most readily achieved through the testing of clearly formulated hypotheses.

Economic success occurs when monetary or labour expenditures are reduced with continuing control of the introduced weed. This is described in more detail below.

Political success results from communication of biologists and land managers with stakeholders, customers, and federal and local government representatives. The goal here is often to achieve appropriate levels of funding or legislation to facilitate future programs.

Social success is achieved through increased awareness among land managers and members of the public with a vested interest in public lands. Often limited budgets result in little money being spent on maintaining reduced weed densities or fighting weeds before they become a serious problem. Social success can be measured by a change in the perceptions of the weed problem and how to deal with it.

Legal success requires drawing up legislation and enforcing it to prevent the introduction of invasive species, to mandate effective control programs, and to assess appropriate penalties for failure to comply to existing laws.

To achieve all of these levels of success in a biological control program is an enormous challenge. We next consider in more detail cases in which the biological and economic successes of biological control have been achieved.

Quantifying biological success

Perhaps the greatest weakness of biological control has been the failure to quantify success adequately and to monitor programs effectively. Thus, whether a program is successful is often a crude value judgment. The

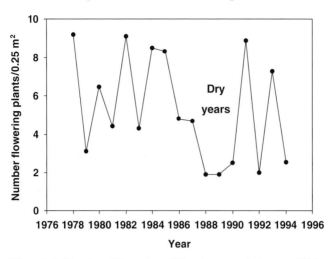

Figure 7.3. Density of flowering diffuse knapweed plants at White Lake, British Columbia, measured in August of each year in thirty 0.25 m² quadrats spaced at 10 paces along two arbitrarily placed transects. (J. Myers, unpublished data.)

most extensive review of biological control of weed programs is 'Julien's Catalogue'. There have now been three editions of this compendium of cases of biological weed control, the most recent having been published in 1998 (Julien and Griffiths 1998). While this has been an extremely useful document, it is based on varying and often subjective evaluations. This weakness carries over to other analyses based on data from the catalogue. An example of how errors can enter this database is a case where the decline of a weed, *Centaurea diffusa*, was associated with a period of drought (Figure 7.3). This population decline is listed in Julien's Catalogue as a 'success', and the status is unlikely to be reversed, even following a resurgence of weed density with the return of more typical rainfall patterns.

McFadyen (1998) recommends using the definitions for biological success originally proposed by Hoffman: **complete success** is when no other control is required, **substantial success** is when further control procedures are required but at reduced levels, and **negligible success** is when control still depends on other measures even if agents are inflicting damage to the plants. A problem with measuring success based on the use of other control measures is that in many cases chemical or physical controls are not practical or economically feasible. Thus they are not used. In addition, one must distinguish reduced control from just giving up or because funds have run out. It is ideal to use a quantitative assessment of plant density as an indicator of biological control success followed by experimental removal of agents to evaluate their impacts (e.g. McEvoy and Coombs 1999). Unfortunately these types of evaluations are rare. The *Proceedings of the Xth International Symposium on the Biological Control of Weeds* contains 21 papers or abstracts presented in a session on 'impact studies' and only one contained quantitative data on density reduction of the target weed. This one study reported the reduction of leafy spurge, *Euphorbia esula*, from approximately 274 stems m^{-2} to 1–14 stems m^{-2} caused by the beetle *Aphthona nigriscutis* (Stromme *et al.* 2000). The other 20 papers were on establishment of agents. Methods for quantifying plant populations are described in the Appendix. It is both surprising and regrettable that these methods have been so rarely used to evaluate biological control programs.

Often, the evaluation of success is based on the establishment and the level of damage caused by introduced agents rather than on the reduction of plant density. It is usually easier to count the number of insects found in flower heads or in the roots of attacked plants than it is to relate this attack to plant density. Reviews of biological control programs tend to report an establishment rate for introduced agents close to 60% with approximately 10–35% of the established species being considered to be involved in control (Crawley 1989, Denoth *et al.* 2002). This gives an overall success rate of approximately 6–20% per species introduction. Some claims of success of projects are higher, particularly if the categories of complete and substantial success are combined (McFadyen 1998, 2000). For example, in South Africa six of 23 targeted weeds are considered to be successfully controlled, with an additional 13 being under substantial control, for an overall success rate of 83%. In Hawaii seven of 21 weeds (33%) are considered to be successfully controlled, with another three being substantially controlled for a total of 50% success (McFadyen 1998). In Australia, 12 of 15 (80%) completed control programs were considered to be completely successful, and of 21 ongoing programs, four achieved complete control and three substantial control (33%) (McFadyen 2000). Fowler (2000) claims a biological control success rate of 80% for Mauritius and 83% for New Zealand. Based on these evaluations success rates are considerably higher than 20%. In her summary of successful biological control projects, McFadyen (2000) lists 41 species of weeds successfully controlled somewhere in the world using insects (Table 7.1). These involved the introduction of 183 species for which 74 (40%) were considered to have contributed to success. Success has

Table 7.1. *Examples of successful biological control of weeds using insects. From McFadyen (2000) based on data from Olckers and Hill (1999), Briese (1999), and Julien and Griffiths (1998)*

Weed species	Countries
Ageratina adenophora	Hawaii
Ageratina riparia	Hawaii
Carduus acanthoides	USA
Carduus nutans	USA, Canada
Carduus tenuiflorus	USA
Centaurea diffusa	USA, Canada
Centaurea maculosa	USA, Canada
Chondrilla juncea	Australia, USA
Chromolaena odorata	Guam, Ghana, Indonesia, Marianas
Senecio jacobaea	Australia, Canada, NZ, USA
Xanthium strumarium	Australia
Hypericum perforatum	Canada, Chile, USA, South Africa
Cordia curassavica	Malaysia, Mauritius, Sri Lanka
Euphorbia esula	Canada, USA
Sesbania punicea	S. Africa
Hydrilla verticillata	USA
Lythrum salicaria	USA, Canada
Sida acuta	Australia
Clidemia hirta	Fiji, Hawaii, Palau
Acacia saligna	S. Africa
Mimosa invisa	Australia, Cook Is., Micronesia, PNG
Emex australis	Hawaii
Tribulus cistoides	Hawaii, PNG, West Indies
Tribulus terrestris	Hawaii, USA
Alternanthera philoxeroides	Australia, China, NZ, USA
Pistia stratiotes	Australia, Botswana, Ghana, PNG, S. Africa, Sri Lanka, Zambia, Zimbabwe
Salvinia molesta	Australia, Fiji, Ghana, India, S. Africa, Sri Lanka, Zambia, Zimbabwe
Eichhornia crassipes	Australia, Benin, India, Indonesia, Nigeria, PNG, S. Africa, Thailand, Uganda, USA, Zimbabwe
Harrisia martinii	Australia, S. Africa
Opuntia aurantiaca	Australia, S. Africa
Opuntia elatior	India, Indonesia
Opuntia ficus-indica	Hawaii, S. Africa
Opuntia imbricata	Australia, S. Africa
Opuntia leptocaulis	S. Africa
Opuntia littoralis	USA
Opuntia oricola	USA

Table 7.1. (*cont.*)

Weed species	Countries
Opuntia streptacantha	Australia
Opuntia stricta	Australia, India, New Caledonia, Sri Lanka
Opuntia triacantha	West Indies
Opuntia tuna	Mauritius

been achieved with terrestrial weeds as well as with aquatic weeds such as *Salvinia* (Box 7.3). The important message is that although success is not a certain outcome of biological control programs, many programs are successful. Often biological control is the only technique available for reducing the density of an introduced plant. Eradication, chemical control or cultural control are just not options for many introduced weeds.

Box 7.3 · *Biological control of the floating fern,* Salvinia molesta.

Salvinia, an aquatic plant native to South America, was first introduced to Sri Lanka in 1939 by way of the Botany Department at the University of Colombo (Room 1990). From there it was spread by humans to Africa, India, southeast Asia and Australasia where it became a serious problem in lakes, canals, rice paddies, and irrigation channels. *Salvinia* is still being introduced to Canada as an aquarium plant. At the beginning of the biological control program in 1972, *Salvinia* was misidentified as *S. auriculata*, but later it was correctly identified as *S. molesta*. The geographic origin of this species was not known until it was discovered in Brazil in 1978. The species is sterile and genetically uniform, although it is morphologically variable. Plants are colonies of ramets held together by branching rhizomes. Room suggests that in the early 1980s *Salvinia* was probably the world's largest single genet covering 2000 km^2 and weighing more than 20 million tonnes. In Grime's scheme (Chapter 3) *Salvinia* is a competitor with a 'general purpose genotype'. It is relatively rare in its native range.

Three species of herbivores were found attacking *S. auriculata* and were initially screened for biological control release. These are the weevil *Cyrtobagous singularis*, the moth *Samea multiplicalis*, and the grasshopper *Paulinia acuminata*. Weevils collected from *S. molesta* were first released in Queensland, Australia, in 1980 and in less than a year had increased to 100 million individuals and destroyed 30 000 tonnes of *Salvinia*. The moth was also released and, while it established, it had no apparent impact. Careful attention to the beetles showed that the species collected from *S. molesta* was not *C. singularis* but a new species, *C. salviniae*.

An interesting twist to the generally positive story of the biological control of *Salvinia* occurred in New Guinea. This weevil that had been so successful in Australia became established following release in New Guinea in 1982. However, it failed to increase here because plants were too low in nitrogen to promote a rapid weevil buildup. Not until fertilizer was used to increase the nitrogen levels of the plants did weevil densities increase and *Salvinia* populations decline. After fertilization this system became self-sustaining as dead plant material increased nitrogen levels in the lake. Fertilizing weeds may not seem a rational thing to do, but in this system it was very effective. This is a good example of integrated weed control (Chapter 9).

Cyrtobagous singularis was never successful as a control agent while the congener *C. salviniae* was. Some reasons for this are that *C. salviniae* causes more damage to the plants per capita by destroying more buds as adults and killing more rhizomes as larvae (Sands and Schotz 1985). *C. salviniae* is also able to tolerate higher densities than *C. singularis*. In this example the insect species that coevolved with the host plant was more effective in controlling it than was a weevil that originated from another species of *Salvinia*.

Once attacked by weevils, the large mats of *Salvinia* break up and sink, or are carried by wind or water currents and packed into temporary aggregations. At very low densities only small ramets persist among fringing vegetation. Beetles must have high searching efficiency to persist at this low-density equilibrium.

Cost effectiveness

Calculating the cost effectiveness of biological control of weeds is tricky. Estimates of costs can be made in terms of person hours required to carry out the exploration for potential agents, research on host specificity and possibly evaluating efficacy of control agents, preparation of the applications for importation and release permits, rearing and release of agents, and monitoring. A program for a new weed will cost between US$200 000 and $500 000 per year over 5 to 15 years for a total of $3–8 million (McFadyen 2000).

Evaluating the costs of an introduced weed is more difficult. In some situations the costs of weeds can be estimated by their impacts on products or productivity. For example, the cost of Noogoora burr, *Xanthium strumarium*, can be estimated from the cost to the Australian wool industry of the loss due to burry wool (Chippendale 1995). In the early 1980s a rust, *Puccinia xanthii*, that had been considered but rejected for release on Noogoora burr, was found to be devastating plants in Queensland (Chapter 6). In addition another biological control agent, the lepidopteran *Epiblema strenuana*, originally released for another weed species, was found to be attacking Noogoora burr. Together these agents reduced Noogoora burr and the loss of wool from contamination by burrs. The average benefit between 1982 and 1991 was calculated as A$837 000 per year. Projecting this figure into the future gives a very large benefit. However, the total cost of the biological control program was estimated to be around A$7.2 million.

Problems associated with evaluating the economics of biological control were outlined by Greer (1995). Cost–benefit analyses allow the monetary comparison of projects with different costs, annual returns, time horizons and probabilities of success. Some of the market effects of introducing a biological control agent are changes in (1) quantity and quality of production, (2) resources needed for production, (3) market prices, if increases in production are achieved, (4) benefits generated by the weed, e.g. nectar sources, and (5) resources needed to implement research. Some non-market effects are those on (1) native flora and fauna, (2) aesthetic values, and (3) on the physical environment.

Biological control also involves much uncertainty about: (1) the establishment and dispersal of the agent, (2) population responses of the agent in the new environment, (3) impact on the target, (4) impacts of plant competitors on the response of the target, and (5) impact of the climate on the agent. The use of contingent valuations is one way to assess the value of a project. With this technique the value is determined as the amount of money a person is willing to pay to find a solution to a problem, in this case an introduced weed. This technique was used to evaluate research into the biological control of *Clematis vitalba* in New Zealand

(Greer 1995). The average respondent was willing to pay NZ\$46.37, and therefore it was estimated that New Zealand society as a whole was willing to pay \$44–111 million for a small chance of controlling *C. vitalba* using biological control. While this value suggests that citizens would be willing to pay for control of one weed, it is unlikely that they would be so generous when it came to all of the other introduced weeds in New Zealand.

An economic evaluation of the control of skeleton weed, *Chondrilla juncea*, indicated a benefit–cost ratio of 112 to 1 (McFadyen 2000). A difficulty that can arise in these analyses, however, is that introduced plants may have positive value for some individuals, such as bee keepers, while they have negative value for others, such as farmers. The European weed *Echium plantagineum* has two common names in Australia, Patterson's curse and Salvation Jane, indicating that it has different impacts in different agricultural systems. Controversy arising following initial biological control attempts led to the Biological Control Act of 1984 in Australia, which for the first time provided legislation for the release of biological control agents (Cullen and Delfosse 1985). In this evaluation, the benefit of control to farmers was calculated to be A\$30 million per annum. This is measured against A\$2 million losses to apiarists and to farmers in semi-arid areas where *Echium* is used as forage (Cullen and Whitten 1995). A similar evaluation of blackberry, *Rubus fruticosus*, was A\$40 million benefit and A\$0.7 loss. An evaluation of *Salvinia molesta* in Sri Lanka was based on 'hours worked', a unit used in aid projects. With a medium estimate of the cost of biological control and a discount rate of 10% (value in the future as compared to the present), the estimated benefit was 62 million hours for a return ratio of 1675 to 1 (Cullen and Whitten 1995). The value of controlling this water weed is particularly high because of its influence on the use of rivers for transport, fisheries, irrigation and water supply. Biological control programs for *Salvinia* in South Africa achieved a benefit of \$276 million in weed control costs by 1998 (Olckers *et al.* 1998 cited in McFadyen 2000). Successful programs apparently have high benefit to cost ratios. This will not, however, be the case for unsuccessful programs. These are difficult to evaluate since it is not clear if the effort was insufficient to achieve control, or if the biology of the target weed is such that biological control is not achievable. Programs that have not been successful are not discussed in these analyses. Perhaps the key metric for biological control is the sum of the benefits divided by the sum of the costs for all successful and unsuccessful programs. Once an infrastructure is in place, the costs of additional programs are lower.

Remembering success

The glory of successful biological control is often short-lived and people quickly forget how annoying or devastating previous weed infestations were. McFadyen (2000) describes how water hyacinth, *Eichhornia crassipes*, was only perceived as a problem in Brisbane, Australia, when there were floods that washed dense mats downstream, blocking ferries and threatening bridges. After the introduction of the weevils *Neochetina eichhorniae* in 1975 and *N. bruchi* in 1990, water hyacinth in the upper reaches was controlled and floods now no longer bring down floating mats of vegetation. However, because the problem was previously sporadic, almost no one recognizes the success. For this reason it is important to record successes photographically to maintain public support for future programs.

Attitudes towards weeds can change as well. St John's wort, *Hypericum perforatum*, is one of the classical cases of biological control success. However, the recent popularity of St John's wort as a herbal remedy for depression has resulted in it being cultivated in some areas where the control agents still persist. Those trying to grow *Hypericum* probably do not remember or care that it was once a serious pasture weed in the same area.

Can we predict successful agents and vulnerable plants?

Are certain plant types more susceptible to biological control?

To keep the ratio of introduced agents to target weeds as low as possible and still to achieve success we would like to be able to predict which weeds are most likely to be vulnerable to biological control and which agents are most likely to be successful. Burdon and Marshall (1981) proposed that asexually reproducing plants were more easily controlled than sexually reproducing weeds. They suggested that this was because clonal plants had less genetic variation and therefore were less able to resist attack. Chaboudez and Sheppard (1995) reanalyzed this relationship by considering changes in the status of the species in biological control projects originally reviewed by Burdon and Marshall, and then by considering other studies for which new data were available. They found that the success of control was independent of the mode of plant reproduction, although complete control of sexual plants tended to be less frequent than that of asexual plants. A factor that might influence these results is that annual plants tend not to reproduce asexually. Temperate species of annual plants have rarely been successfully controlled. Straw and Sheppard (1995) compared the success of biological control of different

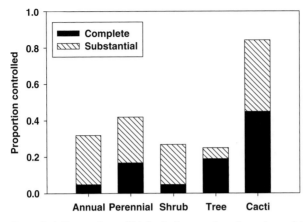

Figure 7.4. Proportion of biological control projects that achieved substantial and complete control. Annual herbs (total projects = 22), biennial and perennial herbs (36), shrubs (21), trees (8), and cacti (20). Data from Straw and Sheppard (1995).

plant types (Figure 7.4) and found no significant variation other than that the control of cacti was higher than of other plant types. Crawley (1990b) also reviewed the relationship of plant life histories to biological success and reported that poor success had been achieved with plants having a rhizomatous, perennial growth form, high powers of regrowth, and low quality as food for insects. It is of interest, however, that these are all characteristics of *Salvinia*, the target of a recent highly successful biological control program (Box 7.3). Plants more likely to be controlled in Crawley's review were those with genetic uniformity, a lack of perennation or dormancy, and those susceptible to secondary infection.

No strong relationship is apparent between the biology of plants and their potential for successful biological control. McFadyen (2000) warns that unjustified classification of certain plant types as being less susceptible to biological control could lead to decisions against initiating a program that could in fact be successful. The strongest general pattern to emerge is that cacti are particularly sensitive to biological control. This is not encouraging for the situation of cacti in North and Central America following the establishment of the *Cactoblastis* moth in Florida (Box 7.2).

Are certain plants more suitable for biological control?

As well as judging the **susceptibility** of plant species for biological control, choosing those that are **suitable** helps to prioritize efforts in dealing with introduced weeds. Peschken and McClay (1995) developed a rank-ing system based on both economic and biological criteria as a guide to researchers, managers and administrators. The economic criteria include the following: the cost of the weed, the extent of the infested area, the potential for extensive spread, the toxicity of the weed, available means for control, and potential conflicts of interest. The biological criteria include whether the species reproduces asexually, its geographic distri-bution, relative abundance, control elsewhere, suitable available agents, habitat stability, and the occurrence of closely related native species or ornamentals in the area of release. Peschken and McClay give different values to these characteristics and a weighted score to determine suitable biological control targets. While these weightings may be arbitrary, they are based on the experience of biological control practitioners and can be of assistance to managers.

Can we predict what will be a successful biological control agent?

How many agents are necessary for success?

Biological control of invasive weeds requires the release of additional ex-otic species. Because unexpected non-target impacts of these introduced agents may occur, it is important that the number of species imported is kept to the minimum necessary for success. Two views exist about how biological control agents act to reduce host plant density, and these relate to how many should be introduced. Some view the impacts of biological control agents on the host plant as being cumulative (Harris 1981, 1984). In this scenario, host plant stress increases with each additional species of herbivore introduced, and with the attack of additional parts of the plant. When sufficient stress to the target plants is achieved, population dens-ities will decline (Figure 7.5). This is the cumulative stress hypothesis. For example, Malecki *et al.* (1993) encouraged the release of two species of leaf feeding beetles, a root feeding weevil and a flower feeding weevil on purple loosestrife, *Lythrum salicaria*, under the assumption that simul-taneous attack of these different plant parts would speed up and produce more effective control.

An alternative view is that only a few species of natural enemy have the potential to successfully control host plant density, after they are re-leased from their own competitors and natural enemies. If single species of agents are frequently successful in controlling host plant density, the cumulative stress model will not be supported. Myers (1985) proposed

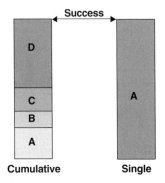

Figure 7.5. Two models for successful biological control of weeds are diagramed. In the first, cumulative stress from multiple agents is required for successful control (the cumulative model). In the second, single agents can effectively reduce host plant density (single model).

that the introduction of biological control agents is like a lottery in which the chance of introducing an effective agent is increased by introducing more agents. However, if one could predict what might be an effective agent, the number of releases could be reduced.

Denoth *et al.* (2002) reviewed successful biological control of weeds programs and asked how many of the introduced agents were considered to have contributed to the control of the target weed. In slightly over half (54%) of the successful projects in which more than one species of agent was released, a single species was considered to have been responsible for the success (Figure 7.6). Therefore, cumulative stress is not necessary for successful biological control.

While single biological control species can control target weeds, in some situations different species may be needed to attack plants in different climatic regions. Therefore, over a landscape involving north and south facing slopes, or wetter or drier conditions, a combination of species may be better. An example here is the control of leafy spurge by *Aphthona* species that differ in their preferred habitats (Anderson *et al.* 2000).

Another project for which multiple agents are thought to be important is with the control of *Sesbania punicea* in South Africa. Hoffman and Moran (1998) studied the impact of three weevil species, one that destroys flower buds, one that destroys developing seeds and one that bores in the trunk and stems of this moderately long lived tree, often killing them. They concluded that, although the decline of the weed over 10 years of observation was most evident in the presence of the boring beetle, the

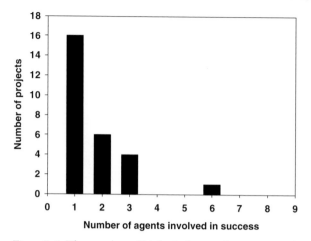

Figure 7.6. The number of biological control agents considered by practitioners to be involved in the successful control of target plants in programs involving the introduction of more than one species. In this survey a total of 153 agent species were introduced of which 53 were considered to contribute to success. Data from Denoth *et al.* (2002).

other beetle species also contributed to population decline. The population declines were significant in most sites with the beetle that kills trees, but the strongest population declines occurred in plots with all three biological control agents (Figure 7.7). Hoffman and Moran (1995) consider the seed predator to be a complementary agent in that it reduces seed production of *S. punicea* by 98% and curtails the rate of spread of the plant into uninvaded habitats and/or those cleared by mechanical means.

However, there are costs associated with the introduction of multiple species, a portion of which may be uneffective. One of these costs is that species might interfere with each other. Establishment of agents in biological control programs was not related to the number of species released (Denoth *et al.* 2002), but it is difficult to believe that the impact of multiple agents is always additive. Other costs are those associated with screening, release and monitoring of multiple species. Focusing resources on one or two agents might allow more resources for monitoring and evaluation. Therefore, given that in many cases a single species can control a weed, a conservative approach should be taken to multiple releases.

In summary, multiple agents can have value in some biological control programs but, in many, one or two agents are sufficient for effective control. Often programs involve the release of three to more than 20 agents. The number of species released could be reduced if successful agents

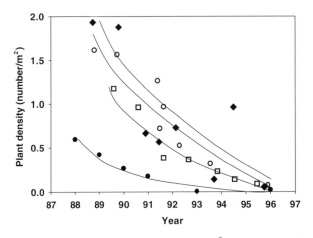

Figure 7.7. The mean density (number m^{-2}) of mature *Sesbania punicea* plants attacked by two species of flower feeding weevils and a boring weevil capable of killing the plant. Data are from four sites containing all three species. After Hoffman and Moran (1998).

could be identified. By looking at successful agents, we may be able to identify characteristics that could be applied to the choice of agents in future programs.

Selecting the right agent

Several schemes have been proposed for identifying good biological control agents based on the type of damage, host specificity, geographic distribution and the evolutionary relationship between the agent and the host (Harris 1974, Goeden 1983, Hokkanen and Pimentel 1984, 1989, Goeden and Kok 1986). Wapshere (1985) compared Harris's method for selecting agents based on the attributes of the agent and the weed to his own system based on observations of the effect of the agents in their native range. He considers 12 criteria (outlined in Table 7.2) and recommends an ecoclimatic approach in which the impacts of potential agents are evaluated in the native habitat in situations as close as possible to the infested region.

When several competing agents are being compared, Wapshere recommends, as did Zwölfer (1973), that the weakest competitor should be chosen. Species that are poor competitors or that have a high level of predation or parasitization in the native habitat are predicted to increase rapidly when released into a new environment. Wapshere suggests that

Table 7.2. *Characteristics of potential biological control agents scored by Harris (1974) and critiqued by Wapshere (1985)*

Characteristic	Harris score	Wapshere comments
Host specificity	Oligophagy better	Not always
Direct damage	Leaf miners, gall-formers, defoliators low score	At high densities can destroy plants
Indirect damage	Not as useful as direct damage	Less useful if it depends on another organism (fungus) to be introduced
Phenology of attack	Continuous attack gets high score	Short attack of some species more damaging
Number of generations	Multivoltine species high score	In some climates univoltine species better
Number of progeny	High fecundity high score	May reflect the mortality risk of early stages and not good predictor of successful control
Intrinsic mortality factors	If high in native habitat agent will excel without these in new habitat	Not easily estimated in the native habitat
Feeding behavior	Territorial behavior or cannibalism low score	Being able to build to high density is good
Compatibility	Competitors for same feeding niche is bad	Has greater value later in program
Distribution	Widespread agent more adaptable	May be less well adapted to a particular environment
Evidence of effectiveness	Success elsewhere gives high score	Should rank higher than other characteristics
Size of agent	Large good	Small agents can reach high densities and be damaging

observations should be made on the impacts of potential agents in dense populations of weeds in the native habitat either in fortuitously created situations or in successional stages similar to the ecoclimatically equivalent infested areas. Agents that respond to and suppress weed outbreaks in the native habitat are likely to be successful biological control agents. These ideas build on the recommendation of Force (1972) that the best biological control agents will be those that are not dominant in undisturbed habitats, are widely spread geographically, and are capable of exploiting high host density associated with disturbance or newly available habitats.

Are new associations of plants and insects more likely successful?

Plants are expected to adapt to their natural enemies. Hokkanen and Pimentel (1984, 1989) therefore suggested that if this is the case, biological control should be more successful if agents are chosen that represent new associations with the target weed, i.e. agents to which the plants have not adapted. Ehler (1995) described four categories of pest–enemy associations: (1) new associations, (2) recent associations, (3) quasi-old associations (50 to 100 years after the target pest was introduced to a new area), and (4) old associations. In these associations one might expect a continuum of adaptation from weed–insect associations that are tightly coadapted in (4) above, to those in which the weed is such a poor host for the insect that levels of attack are low, as in (1) above. A new association could be based on geographical isolation or on the agent being maintained at low density by high mortality or competition in the native habitat and, thus, having a greater impact on the host when the species are brought together in a new environment. If the host plant species had not adapted to the agent it should be more susceptible to the attack. Dennill and Moran (1989) found that the associations between insects and host plants among insect–crop interactions in South Africa supported the prediction of stronger impacts of insects on plants in new associations. This does not mean that old associations should be excluded when looking for biological control agents since there are successful programs involving these including the successful control of *Salvinia* (Box 7.3).

These ideas have recently been summarized in the 'biological tolerance hypothesis' as a guideline for selecting control agents (Myers 2001) (Figure 7.8). This proposes that plants will evolve tolerance to common natural enemies, but will have little resistance to natural enemies that are rare in the native habitat. Release of the latter species without their

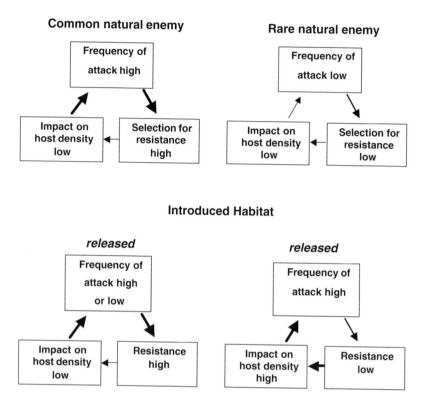

Figure 7.8. The 'biological tolerance hypothesis' predicts that natural enemies that are rare in the native environment will not select for increased tolerance or resistance of their host plants. However, when these agents are released without their own natural enemies as biological control agents, they will thrive and reduce the density of susceptible host plants. The width of the arrows indicates the strength of interactions. After Myers (2001).

competitors and natural enemies can result in rapid population increase and high levels of attack of the susceptible host plants. These are the conditions associated with successful biological control. While these ideas are speculative, they suggest that it might be better to only look for rare rather than the most common species of natural enemies in the native habitat. Species released from their own competitors, predators, and parasitoids in a new habitat may be very effective.

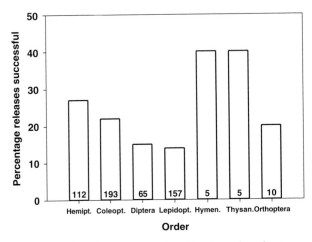

Figure 7.9. The relative success (combination of moderate, marked or complete) of biological control agents in different insect orders. Numbers of cases are given on the bars. Data from Crawley (1990b).

Some successful control agents are known to have been rare in their native habitats (Myers 2001), but it is often difficult to find information on the population status of biological control agents before their release. An excellent approach to selecting biological control agents is to create experimentally high densities of target weeds in native habitats and to measure the response of natural enemies.

Historical perspectives – using the past to predict the future

Another approach to finding successful biological control agents is to look at the types of agents that have been most successful in the past. Crawley (1990b) reviewed biological control programs and found that only a few insect orders have been involved in a large number of introductions. Of these, Hemiptera (sucking insects) and Coleoptera (beetles) have been most successful in biological control (Figure 7.9). The 43% success of cochineal insects, *Dactylopius* spp., in controlling cactus, and the large number of programs against cactus world-wide, 45 by 1985, will have contributed to the high success rate of Hemiptera. In contrast, the gall-forming Tephritid flies have been introduced in 25 biological control programs without success (Myers and Ware 2002). This contributes to the low success of Diptera. Crawley (1990b) proposed that both establishment and success are related to a high rate of increase, long-lived adults,

Table 7.3. *Factors found to have influenced the impacts of agents in different programs and to cause variable success among projects*

Variable factors influencing the impact of control agents
Adaptation to form of weed
Temperature effects on agent or on agent and weed
Moisture effects on agent or on agent and weed
Other environmental conditions – sun/shade, cultivation, refuges
Predation/parasitism on agent
Compensation by plant for attack
Competition on weed from other plants
Life cycle adaptation

Adapted from Cullen (1995).

multiple generations per season, and small individual size. A review of South African weed control programs found stem borers to be the most successful agents followed by sap suckers (Olckers and Hill 1999). It is not easy to identify successful biological control agents, but those that kill plants and have prolonged attack do seem to be more effective.

Cullen (1995) took another approach to identifying good biological control agents and asked why agents work in some locations but not others. He evaluated 25 projects that varied in their success around the world, and identified factors thought to be responsible for this variation in success (Table 7.3). Environmental factors and predators at release sites were particularly important in determining the variation in success. In this comparison, conditions of the site were more relevant than characteristics of the agents.

The literature on biological control is a rich source of information on the interactions between the environment, plants, and insects. It is not surprising that no single factor emerges to identify the type of agent with the highest probability of success although sucking insects and beetles have been the most successful. Different factors combine to make agents work well in many, some or no situations. The study of biology can never be left out of biological control.

Do seed predators make good biological control agents?

One of the most obvious impacts of insects that do not kill their host plants is reducing seed production. Some insects attack buds, flowers and seeds directly while others reduce the vigor of the plant and this is reflected in their size and seed production. In Chapters 5 and 8 we discuss seed limitation and conclude that many plant species are not seed limited, that plants are often able to compensate for seed reduction by better survival of seedlings, and that seed reductions of 95–99+% are required to have an impact on host plant density in many studies.

Seed predators are frequently employed as biological control agents because they are relatively easy to find, particularly if they develop in the seed heads. This may be the reason why they are often introduced. For example, in the control program targeted against the yellow star thistle, *Centaurea solstitialis*, in California, six species of seed predators have been introduced (Pitcairn *et al.* 1999). At some sites, seed and seedling densities have been reduced for several years, but no reduction of plant density has been reported for this annual plant. Would one species of seed predator have been as effective as six?

One example in which biological control success has been attributed to a seed feeder is the control of nodding thistle, *Carduus nutans*, in North America by the weevil *Rhinocyllus conicus* (Kok and Surles 1975, Harris 1984). This control agent also has an impact on rare, native thistles in North America (Louda 1999). A comparison of nodding thistle biology with that of the related weed diffuse knapweed, *Centaurea diffusa*, may indicate why a seed feeder is successful in one case but not the other. Nodding thistle is an annual or short-lived perennial and depends on disturbance for establishment. Once established it can dominate a site, at least temporarily (Wardle *et al.* 1995). In comparison, knapweed is a short-lived (2–5 years) perennial and is able to invade undisturbed sites (Berube and Myers 1982). Knapweed has very high summer survival of the rosette stage (90–95%), while rosette survival of nodding thistle is low (10–30%) (Sheppard *et al.* 1994). Simulation models of knapweed indicate that the only way to reduce weed density would be through an agent that killed rosettes (Myers and Risley 2000) (see Chapter 8). Similarly, Kelly and McCallum (1995) found that density-dependent survival can compensate for seed loss in nodding thistle, but high rosette mortality reduces the situations in which this will occur. This example suggests that seed predators may be effective in reducing the density of host plants that are poor competitors and that have low rosette survival.

As mentioned above, seed predators may contribute to biological control by reducing the seed bank or the spread of a weed. If invasive plants spread by diffusion (see Chapter 8), then reducing seed production may be effective. However, if plants are spread by being carried on equipment, animals, in hay or as tumbling plants, seed reduction will be less effective on lowering the rate of invasion. For woody plants seed predators may be useful in three situations: (1) when the plants are still spreading, (2) when there is a conflict of interest and existing stands of trees are of some benefit but spread is undesirable, and (3) in conjunction with herbicide use (Hartley 1985).

Weedy plants such as thistles have many species of seed predators and the relationship between species packing and resource utilization in the flower head varies among plant species (Zwölfer 1985). The existence of many seed predators may be a hint that weedy plants are well adapted for losses of high proportions of their seeds to these predators. The value of introducing seed predators and particularly in introducing several species of seed predators in a biological control program should be carefully evaluated.

Is biological control safe?

The most common question asked by the public in a discussion of bio-logical weed control is, 'Will the insect start attacking my garden plants when the weed is controlled?' Testing the host specificity of potential biological control agents has been an essential part of programs since the 1920s. Kluge (2000) summarizes the status of biological control of weeds and points out that over 100 years approximately 1049 deliberate releases, resulting in 603 successful establishments of 259 insect species on 111 plant species, have taken place in 70 countries. In addition, there exist 169 records of accidental releases of 63 species of insects on 45 weed species in 51 countries. No biological control agents have become agri-culturally important pests, nor have they shifted to very different host plants (McFadyen 1998). The host plant shifts that have occurred have almost all been to plant species that are congeneric with the target weed (Pemberton 2000b).

However, some biological control agents do have relatively broad host plant ranges and in some cases non-target interactions have occurred. Two situations that have received recent attention are the attack of native cacti by *Cactoblastis* moth in North America (Box 7.1) and the attack of native thistles by the seed feeding weevil, *Rhinocyllus conicus* (Louda *et al.* 1997). Initially the *R. conicus* was introduced from Europe to North America for the control of *Carduus* thistles. At this time it was known to feed on some *Cirsium* spp. native to North America. In their pre-release report Zwölfer and Harris (1984) did not consider an expansion of the

Table 7.4. *Suggestions for expanding and improving procedures prior to the introduction of biological agents to reduce the risk of introducing ineffective species or those with the potential to cause ecological impacts on non-target plant species*

Suggested improvements	Rationale
Enhanced problem definition	Preliminary cost–benefit analysis needed
Expanded ecological criteria	Flowering phenology of native species should be tested to determine threat from insect oviposition
Quantification of ecological interactions	Determine which species could be used by the agent and what would be the outcome
Better assessment of potential efficacy	Good probability of effectiveness in many conditions before release
Evidence on alternatives	Optimization of management strategies
Peer review and public input	Workshops to develop criteria and evaluate evidence
Formal post-release monitoring	Guidelines for further distribution needed
Decision making process	Use biological control where most needed

Based on Louda (2000).

weevil host range to native *Cirsium* thistles to be a threat because *R. conicus* preferred *Carduus* spp., they performed better on it, and the low density of the native thistles made them unlikely hosts. However, *Rhinocyllus* is now attacking Platte thistle, *Cirsium canescens*, in Nebraska and has reduced the number of seeds in one population by 80% (Louda 1998). In response to this situation, Louda (2000) reviewed the background on the initial release of *R. conicus* and suggests ways in which lessons learned might be used to improve future biological control programs (Table 7.4). Her suggestions are in agreement with Thomas and Willis (1998) for making biological control ecologically sound but still feasible.

Are Louda's recommendations a useful template for new laws and regulations? The Hazardous Substances and New Organisms Act of New Zealand will achieve some of Louda's recommendations, but these regulations also have the potential to delay progress in biological control (Fowler *et al.* 2000). How then should biological control proceed? As outlined above, selecting agents has depended on (1) assessing the agent in the native range, (2) its suitability based on ease of being reared, host plant specificity, and potential safety, and (3) its potential effectiveness in the area of introduction. McEvoy and Coombs (1999) recommend using a minimum number of control agents by combining top-down control

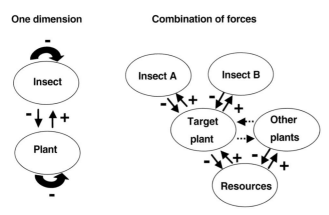

One dimension   Combination of forces

Figure 7.10. Diagrams of one-dimensional control and multi-dimensional control through coordinated manipulation of the biological control agents, plant competition, and disturbance. Positive and negative interactions are indicated by arrows. Solid arrows are direct interactions; dotted arrows indicate indirect interactions. Adapted from McEvoy and Coombs (1999).

of the target plant and bottom-up limitation of resources through increased competition or reduced disturbance (Figure 7.10). In addition, agents that attack life-cycle stages of plants that do not compensate for reduced density will be more effective (those that kill rosettes rather than seedlings) (see Chapter 8). The aim is to use ecological principles to make biological control more effective and efficient.

Conclusions

Biological control is an experimental branch of applied ecology with potentially large economic and social benefits. Successful biological control is often achieved by a single agent, although in some cases other agents contribute to success. Success rates for biological control can be high if sufficient resources are devoted to the project. However, it is important to be parsimonious in the number and types of agents released in programs to reduce the potential for accidental attack of non-target plants, to reduce the possibility of negative interactions among agents, and to conserve resources for monitoring and distribution of agents already approved for release.

There is a rich literature in biological control, much of which is found in the *Proceedings of the International Symposia on Biological Control of Weeds*. These studies have contributed to a better understanding of the

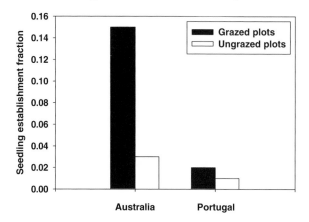

Figure 7.11. A comparison of grazing and seedling establishment of *Echium plantagineum* studied in Australia, the introduced habitat, and Portugal, the native habitat. This suggests that an agent targeting seedling establishment or reducing grazing pressure might reduce the density of *E. plantagineum* in Australia. After Grigulis *et al.* (2001).

interactions between insects, diseases and host plants, although it is still impossible to predict which natural enemies can control their hosts. Studies of plants and natural enemies in both the native and introduced environments should be a necessary prerequisite for biological control programs. They are excellent ways to discover the differences between the mortality of life stages of the plants in the two situations (Figure 7.11) and, thus, the stages to target with natural enemies. Also, studies in the native habitat can reveal which of the potential species are likely to be effective control agents. More quantitative monitoring of control programs will help to make biological control more predictable. It would be a great loss if regulations on the introduction of agents for biological control became overly restrictive. For many invasive weeds biological control will be the only solution.

8 · Modeling invasive plants and their control

Introduction

We would like to be able to predict the dynamics of introduced plant species in different situations, how they might respond to biological control, and how they might spread. Several different types of models have been used to integrate information on the populations of introduced species and their control. These models include (1) simulation models based on individual population units that can vary depending on survival and reproduction functions estimated from field studies and may involve stochasticity, (2) analytical models in which functions derived from simulation models or field measurements are used to describe the population processes, and (3) matrix models based on life table studies. In Chapter 5 we described the most basic aspects of population ecology – birth, immigration, death and emigration – and discuss how life tables could be used to summarize data on the transitions among different life stages. Also we described how the rate of growth, R_0 or λ, of a population could be determined by relating the population density of one generation to that of the next. In this chapter we explore theoretical models of biological control, the use of models to study populations of introduced plant species, and then models of the spread of introduced species. The strengths and weaknesses of different models will be evaluated. A more extensive treatment of models of weed populations can be found in Cousens and Mortimer (1995).

The history of modeling biological control

Models of the biological control of insects were among the first formalizations of interactions between predators or parasitoids and hosts (Thompson 1930, Nicholson 1933). Plants and herbivores can also be described by these interactions. These models were based on the number of hosts (H_t), the number of attacked hosts (H_a), the number of parasitoids

(P_t), the intrinsic rate of natural increase of the host (r), and the conversion rate of hosts into parasitoids (c). Assuming a random distribution of parasitoid attacks, f (the fraction escaping parasitism) $= e^{-aP_t}$, changes in the numbers of hosts and parasitoids can be described as follows:

$$H_{t+1} = e^r (H_t - H_a)$$
$$P_{t+1} = c H_t (1 - f(P_t))$$

where a is the attack rate of the parasitoid.

These early models were very simple representations of the interactions between parasites and hosts and were based on discrete generations (see further discussions in Murdoch 1990 and Begon *et al.* 1996). They sought conditions under which an equilibrium state could be achieved. However, this goal was elusive. The Nicholson and Bailey models and simplified laboratory experiments between parasites and hosts generally result in increasing oscillations between hosts and parasites. Part of the instability of the Nicholson–Bailey models is due to the parasitoids searching for hosts in a random manner. Non-random search, with focus by the parasitoids on high-density host population patches, can stabilize these models under some, but not all, conditions (Hassell and May 1974). Allowing parasitoids to move among host patches within a generation again destabilized the models (Murdoch and Stewart-Oaten 1989). To stabilize these models, exponential growth of the hosts must be replaced with density-dependent growth by including a carrying capacity for hosts in the absence of parasitoids.

The underlying paradigm of the Nicholson–Bailey models is that biological control will reduce the pest to a new low and stable population density. Thus the emphasis was on the biological control agent both reducing host density (limiting factor in Chapter 5, Table 5.1) and regulating the density around this lower equilibrium. However, a general outcome of models is that stability is greater at a higher equilibrium density (see Murdoch 1990 for a review). In addition the ability to limit and regulate hosts may not be general characteristics of biological control agents.

In an attempt to rationalize the instability of the simple models Nicholson suggested that field populations would be divided into subpopulations with a regional pattern of extinctions and re-establishment (see discussion of metapopulations in Chapter 2, Box 2.8). This 'hide and seek' situation was proposed to apply to the successful biological control of prickly pear cactus, *Opuntia inermis* and *O. stricta*, by the *Cactoblastis* moth in Australia. However, field studies (Myers *et al.* 1981) found that

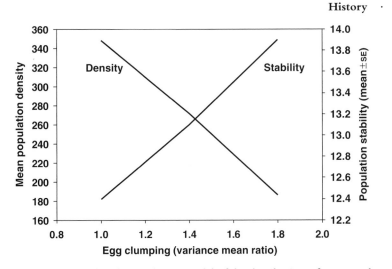

Figure 8.1. Results of a simulation model of the distribution of eggs on plants showing that increasing the contagiousness of egg distribution can increase population stability at the same time as reducing the mean plant population size as predicted by Monro (1967). Data from Myers (1976).

woodland populations of cactus and moths fluctuated regionally rather than locally with fluctuating drought conditions.

Monro (1967) (see picture in Figure 7.2) proposed that stability of insects and host plants could be achieved if insects lay eggs in a contagious distribution so that when populations of the insect herbivores are high, some plants would still have low levels of attack. This idea was formalized in simulation models of differential oviposition choice of insect herbivores that lead to varying degrees of egg clumping on hypothetical plants (Myers 1976). In these models the creation of refuges (unattacked plants) was shown in general to stabilize populations of plants and insects (Figure 8.1) and at the same time to reduce mean plant population density. Tradeoffs between fecundity, egg batch size and distribution, and the ability of larvae to disperse from overcrowded plants were explored with these models. The density and stability of host plant populations were greatest when the number of eggs laid per egg batch was equal to or slightly larger than that which could be supported by the average food plant. This prediction was supported by field observations of cinnabar moth, *Tyria jacobaeae*, introduced as biological agents on tansy ragwort, *Senecio jacobaea*, in Canada (Myers and Campbell 1976).

The factors leading to a clumped distribution of insect eggs on food plants were shown in field studies of the oviposition choices of

Cactoblastis moths for large green cactus plants, and for plants in the vicinity of previously attacked plants. Oviposition choice leads to over-crowding on some preferred plants when insect densities are high and/or plant densities low, but lower densities occur on less preferred plants under these conditions. When insect densities are low, larvae thrive on preferred plants. This unequal distribution of insects promotes population persistence for two reasons: some plants escape attack and will live and reproduce, and, although some insects may run out of food, others will have sufficient amounts (Myers *et al.* 1981). A similar conclusion was drawn by Chesson and Murdoch (1986) for insect parasitoids – when vulnerability to attack is highly skewed in the pest population, stability increases.

Insects and parasitoids received much more attention than plants and insects in the early development of models. However, Caughley and Lawton (1981) applied a Nicholson–Bailey type model to the success-ful biological control of *Opuntia* by *Cactoblastis* in Australia. This model incorporated a carrying capacity for plant density and a factor for in-terference among the herbivores, essentially a carrying capacity to insect density. Their model is described below.

$$dV/dt = r_1 V(1 - V/K) - c_1 H[V/(V + D)]$$
$$dH/dt = r_2 H(1 - JH/V)$$

where V = plant density, K = carrying capacity (5000 plants per acre), H = herbivore density, c_1 = maximum rate of food intake per herbivore (estimated to be large), r = intrinsic rate of increase (r_1 for plant = 2 and r_2 for insect = 3.6), D = inverse of grazing efficiency at low density (ability of moths to find plants), J = number of plants needed to sustain a herbivore at equilibrium (2.23 cactus units per egg stick).

Data for the model were collected in the early stages of the cactus control program by Dodd (1940) and later by Monro (1967). Using these parameters the model results closely mimicked the field observations of a rapid decline in cactus population density and a continuation of populations at a lower equilibrium.

Models of biological control of introduced weeds have been reviewed by Barlow (1999) and Table 8.1 summarizes some of these. How valu-able models and theory have been to biological control still remains controversial. The deterministic models of the Nicholson–Bailey and Lotka–Volterra types are unrealistic and their results do not apply well to

Table 8.1. *Target weeds and outcomes of the models of biological control of weed dynamics reviewed by Barlow (1999)*

Model	Plant	Outcome	Reference
Population simulation	Noogoora burr *Xanthium occidentale*	No long-term dynamics No tests	Martin and Carmajan (1983)
Leslie matrix	Knapweed *Centaurea* sp.	99.5% seed reduction necessary for control	Cloutier and Watson (1989)
Age specific population growth	*Sesbania punicea*	Predicted age structure 4–5 years after introduction of control agents	Hoffman (1990)
Physiological time – within season simulation	Musk thistle *Carduus thoermeri*	Temperature related asynchrony between beetle oviposition and bud development	Smith *et al.* (1984)
STELLA modeling software	*Hydrilla verticillata*	Reproduced plant biomass dynamics	Santha *et al.* (1991)
Photosynthesis-based simulation	Water hyacinth *Eichhornia crassipes*	Predicted plant biomasses not consistent with those observed	Akbay *et al.* (1991)
Simulation model	*Striga hermonthica*	95% of seeds must be destroyed to reduce plant densities by 50%	Smith and Holt (1993)

actual biological systems. Murdoch (1990) warns that all models should be viewed critically, but even simple models may generate useful insights. Murdoch and Briggs (1996) argue that stage structured models can be useful. They propose that realistic models that apply directly to one or several interactions will yield substantial progress. We next review some biologically detailed models of introduced plants and in some cases their control.

Modeling the impact of seed predators

The impact of seed predators on host population dynamics is particularly interesting for two reasons: seed predators are commonly introduced as biological control agents (Chapter 7), and experimental studies indicate that only about half of the plant species studied are seed limited (Chapter 5). For biological control programs, it would be nice to be able to predict if lower seed production would reduce the population density or restrict the spread of an introduced species. This is one of the uses of models of populations of introduced plants.

The simplest model of a plant population can be used to determine the mortality required for population stability when fecundity is high, as is often the case for weeds. If a plant produces on average 300 seeds, 99.7% of these will have to die prior to becoming reproductive plants in order to maintain a stable population (Myers 1995). For the density to decline, mortality will have to be greater than this, but not much. The plant population would double if only 99.5% of these died. In a life table, the mortality occurring at different stages is measured and these are combined over the generation (Chapter 5). Field estimates of the mortality of each life stage will have an error associated with them and these errors are likely to be greater than the difference in mortality between a declining and an increasing population. In addition, the mortality between life history stages will vary from year to year with weather conditions. This simple example indicates that caution is necessary in interpreting population models in terms of predicting the future density of a plant population. However, models are a good way to formalize concepts and integrate data on plant populations. They should, however, include relationships based on the ability of populations to compensate for increased mortality of different life stages or for reduced fecundity.

Several models of introduced plants have focused on the role of seeds in the population dynamics of plants. One of these is the study of Powell (1990a) on diffuse knapweed, which was described as a case study in Chapter 5. This model indicated that diffuse knapweed was well buffered against herbivore attack and agreed with experiments showing that knapweed could compensate for increased plant mortality (Myers *et al.* 1990).

Another model of knapweed

Powell's model was based on a single population of knapweed. To test the robustness of the results, Myers and Risley (2000) developed another model of diffuse knapweed in which the relationships between density and seedling survival and seed production were obtained for five different introduced populations in British Columbia. To this were added the impacts of biological control agents that primarily reduce the seed production of the plants: two species of gall flies, *Urophora affinis* and *U. quadrifasciata*, and a root-boring buprestid beetle, *Sphenoptera jugoslavica*. Gall flies were estimated from field observations to reduce seed production by approximately 8.3 times and seed production of populations attacked by

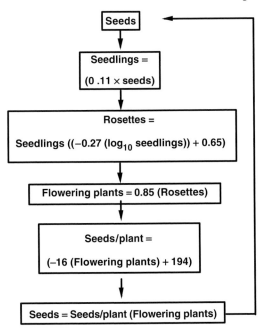

Figure 8.2. Outline of the life stages and functions used in a simulation model of diffuse knapweed introduced to the Okanagan Valley, British Columbia. The outcome of the model was strongly influenced by the relationship between plant density and seed production and to stabilize the model the slope of this relationship was reduced to a lower level than that observed in field populations, i.e. less density dependence than observed. After Myers and Risley (2000).

beetles was 0.74 times that of populations without beetles. Together, these insects can reduce seed production by 95%, a level that might be expected to have measurable impacts on knapweed density. Finally, a hypothetical agent that killed rosette plants was introduced as a mortality agent, and this mortality was set at 50%.

The outline of the model is shown in Figure 8.2 and the results in Figure 8.3. Simulations showed that reducing seed production could in some situations actually increase plant density. Sufficient reduction in seedling density improved survival and high seed production of flowering plants in low-density populations increased reproduction. A hypothetical agent that killed rosettes was shown to have a potentially large impact on host density. Insects that kill diffuse knapweed rosettes have not been found in native populations. This model indicates that there is little hope that seed predators will reduce knapweed density.

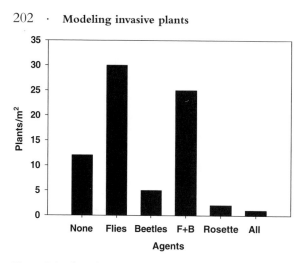

Figure 8.3. Plant density predicted from simulations of the impact of biological control agents on introduced populations of diffuse knapweed in British Columbia. Agents are none, gall flies attacking seed heads, root boring beetles, both flies and beetles (F + B), a hypothetical agent that kills rosettes, and all agents combined. After Myers and Risley (2000).

A hypothetical, stochastic model of seed limitation

Maron and Gardner (2000) developed a theoretical, stochastic simulation model to explore whether reduced fecundity could have population level consequences for plant populations. This model was based on a square grid of 400 cells simulating 1 m² plots that contain plants. The model differed from the knapweed models in that its parameter values were hypothetical, a seed bank was included among the parameters, seed production was not related to plant density, and safe sites for seed germination were incorporated as a variable. Safe sites are places where seeds can germinate. Once these sites are filled, additional seeds are surplus, and therefore consumption of these by herbivores would not influence population density. The availability of safe sites in the real world could vary temporally and spatially with disturbance. Variables of the model included the percentage of years or sites in which recruitment was seed limited, the recruitment of seedlings correlated directly to seed numbers (as compared to site limited), and a set proportion of seeds, generally 5%, that was allowed to germinate in safe sites. Survival of seedlings was related to seedling density using the relationship: $\{(ab)/[a + (b - a) \exp(-\alpha$ seedling number)]\}$ in which a is the minimum seedling mortality (0.7) and b is the maximum seedling mortality (0.95). How realistic these limits are is uncertain. For diffuse knapweed, seedling mortality was 0.2–0.65

at low seedling densities, and therefore field populations of plants might be expected to have a larger range of seedling mortalities than used in this model. The impact of herbivores on seed production in this model was set at 2- and 2.5-fold increases without herbivores. This would be considerably lower than that observed in knapweed populations described above in which a conservative estimate of 8.3-fold difference was used.

In simulations in which recruitment was seed limited (seeds could germinate at all sites which were safe sites for seeds), and there was no seed bank, simulated population sizes were relatively insensitive to differences in seed abundance, although they did increase modestly with greater seed production associated with simulated reduction in herbivore impact. This was due to the density-related seedling survival. Mean population sizes were most strongly influenced in these models by the percentage of years in which recruitment was seed limited rather than site limited. By definition, seed limitation should relate directly to plant density. Herbivores explained 15–17% of the temporal or spatial variance in adult plant population sizes in these models. The simulated impact of herbivores in these models was on seed accumulation in the seed bank. Seed banks buffer populations. Clearly, without field estimates of parameter values and field experiments on the ability of plants to compensate for seed reduction, these simulation results must be interpreted with caution.

Models of Scotch broom

Simulation and analytical model of native populations of broom

Scotch broom, *Cytisus scoparius*, has been the topic of several modeling exercises and these provide interesting comparisons of approaches and results. Rees and Paynter (1997) developed both simulation and analytical models to explore the population dynamics of Scotch broom in Britain. Their simulation model was based on a large number of identical sites, approximately the size of a broom plant, arranged in a 75 × 75 square lattice. Each iteration of the model included the following: (1) some sites become disturbed, (2) seeds germinate following a Poisson distribution, (3) seeds are reduced according to a decay probability, (4) broom plants recruit to disturbed sites in a density-independent manner, but they do not recruit to undisturbed sites, (5) disturbed sites to which broom did not recruit become unsuitable, (6) sites with plants older than a minimum age produced seeds and these could remain in the site or move to eight neighboring sites, (7) older plants senesced and these sites became suitable for broom recruitment, and (8) plants grew older. The simulation is spatially explicit and incorporates local competition, asymmetric competition between seedlings and established plants, local seed dispersal, a seed bank, and an age-structured established plant population.

The results of simulations were evaluated as the equilibrium proportion of sites occupied by broom. Some parameters had little impact on the proportion of occupied sites. These parameters were the probability that a seed becomes a seedling, that a seedling survives the first year, that a seed stays at the parental site, and the decay rate of seeds. However, changing the maximum longevity of plants or the minimum age for reproduction had a strong effect on the proportion of sites occupied by broom. Seed production increased the number of sites occupied over a range of 100 to 1000 seeds, but beyond that there was no further increase. The probability that a site becomes open after plant senescence was directly related to site occupancy and the probability of disturbance had a humped relationship to site occupancy: too little and too much disturbance reduced occupancy.

Earlier experiments on broom by Waloff and Richards (1977) removed insect herbivores with insecticide treatments and these showed that seed production was reduced by 75% by insect herbivores. Rees and Paynter simulated the impact of increased seed production as would occur without insect herbivores. This showed a dramatic increase in the proportion of sites occupied when seed survival was very low (0.01), but not when it was higher ($s = 0.5$ or 0.9). This suggests that biological control agents might be effective in situations of extremely low seedling survival. Seed production data for broom from several exotic populations were used in simulations to explore under what conditions insect herbivores might successfully reduce broom populations. These indicated that in an Australian location, under a eucalyptus forest where broom fecundity is low, insect herbivores could reduce host populations even if seed survival were relatively high, i.e. 0.5 or 0.9. In an open pasture situation, however, in which broom fecundity was higher, the reduction of seeds by insect herbivores would have little impact.

The results of this model show that disturbance is important to broom populations. The authors point out that managers attempting to control broom might try to encourage interspecific competition. The presence of perennial grasses reduces the survival of broom seedlings in pastures. Reducing disturbance could also be beneficial to broom control. However, as we will see in the next section, broom invades quite well without disturbance in some situations (Parker 2000). Rees and Paynter (1997) concluded that a reduction in plant fecundity could reduce broom density when the disturbance rate is high, and plant fecundity and seedling survival low. However, these conditions do not apply to most introduced populations of broom and Paynter *et al.* (1996) concluded that seed predators were unlikely to reduce broom populations in New Zealand. They did suggest that seed predators might slow the spread of the species, but this is not supported by observations in Oregon where the introduced insect *Exapion fuscirostre* reduced seed production by 85%, but did not reduce either broom density or spread (Andres and Coombs 1992).

Matrix models of introduced broom in North America

Broom is an invasive shrub on the west coast of North America and Ingrid Parker's study of two populations of broom, in an urban site and a prairie site, were introduced in Chapter 5 (see Figure 5.10). In her demographic analysis of these populations, Parker (2000) used a projection matrix model to summarize transitions between life stages. The form of this model is $\mathbf{n}(t + 1) = \mathbf{A} \cdot \mathbf{n}(t)$ where $\mathbf{n}(t)$ is a vector of stage abundances at time t, and $\mathbf{A}$ is a matrix of a_{ij} values that describe the contributions of each stage to all other stages at the next time step. The dominant eigenvalue λ of $\mathbf{A}$ represents the asymptotic finite rate of increase at the stable age distribution (see Box 8.1). For this study Parker established permanent plots and classified plants as seeds, seedlings, juveniles, and four size classes of adults (Figure 5.11). Over 3000 individuals were mapped and followed over the study. Fecundity and the probability of transition from one stage to another were determined for these plants. These populations did not have stable age distributions because they were still expanding. Therefore, they did not actually meet the requirements for a matrix model. Altogether Parker developed matrix models for six different sites and within these she distinguished edge and interior sites and for two areas she considered the transitions in two separate years. For these populations values of λ (finite rates of increase) were mostly greater than 1, indicating growing populations. For populations in the centers of the broom distribution λ values were slightly below 1 indicating that they had reached equilibrium. Elasticities measure the proportional change in the different life history transitions and can indicate those most relevant (see Box 8.1). Elasticities were similar across life history transitions in most plots. However, in center plots, elasticities were dominated by what happened to extra large plants. Biomass was a better measure of invasion than density in this study. Simulations of the dynamics of edge plots showed that urban sites took from 46 to 178 years to increase from

Box 8.1 · *Matrix models.*

Life histories of invasive plant species can be presented in a matrix form (Figure 8.4). Age based matrices are referred to as Leslie models and stage based matrices, most frequently used in plant studies, are called Lefkovitch models. Growth, survival, and reproduction transitions are determined from field studies and experiments. The asymptotic growth rate of the population is given by the dominant eigenvalue, λ, of the matrix **A**. The stable stage structure and reproductive values are given by the corresponding right and left eigenvectors, **w** and **v**.

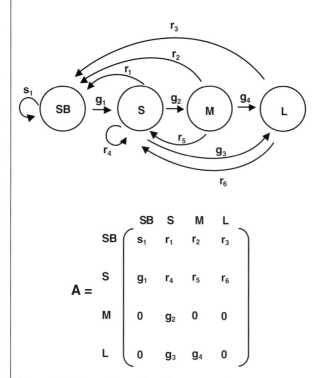

Figure 8.4. Life cycle graph and structure of size-based population projection matrices of *Carduus nutans*. Circles represent stage classes; SB (seed bank), S (small), M (medium) and L (large) plants. Arrows indicate the transitions between classes. Transition labels indicate the probability of individuals at one stage moving or contributing to the next over the transition period of a year; g refers to growth between stages, r_1 to r_3 indicate contributions to the seed bank, r_4 to r_6 indicate seeds that germinate immediately. Dormancy is indicated by s_1. The transition matrix is also shown. After Shea and Kelly (1998).

Projection matrix	SB	S	M	L
SB	0.0382	8.2499	179.4128	503.1428
S	0.1847	1.0906	22.1805	62.1848
M	0.0000	0.0091	0.0000	0.0000
L	0.0000	0.0056	0.0220	0.0000

λ = 2.2142

Elasticity	SB	S	M	L
SB	0.0043	0.1984	0.0177	0.0311
S	0.2472	0.3090	0.0258	0.0453
M	0.0000	0.0448	0.0000	0.0000
L	0.0000	0.0752	0.0012	0.0000

Figure 8.5. Projection matrix and elasticities for a population of nodding thistle, *Carduus nutans*, in New Zealand. After Shea and Kelly (1998).

It may be of interest to know which stage or stages of the life history of a plant population have the greatest impact on variation in population growth. This could be particularly relevant if one were evaluating potential biological control agents that have impacts on different life stages. For example, would a seed predator that reduces fecundity of the plant be more effective than a fungus that kills plants? Perturbation analysis can be used for this and involves the assessment of the impact of small changes in the vital rates on λ. Two types of perturbation analysis are sensitivity analysis and elasticity analysis (Benton and Grant 1999). Sensitivities measure how λ changes with an absolute change in each a_{ij} The sensitivity, s_{ij}, of λ to change in the matrix element a_{ij} of **A** is given by the partial differential, ∂.

$$s_{ij} = \frac{\partial \lambda}{\partial a_{ij}} = \frac{v_i \omega_j}{<W,V>}$$

where $<\ >$ represents the scalar product of the vectors.

Elasticity e_{ij} is a measure of proportional changes in a_{ij} and has the advantage of adding up to 1.

$$e_{ij} = \frac{a_{ij}}{\lambda} \frac{\partial \lambda}{\partial a_{ij}} = \frac{\partial \log \lambda}{\partial \log a_{ij}}$$

The projection matrix of one of the populations of nodding thistle studied by Shea and Kelly (1998) described in Figure 8.4 is shown in Figure 8.5. In this example the three elements of the transition matrix that contribute most to the dominant eigenvalue are those representing the transitions S–S, S–SB, and SB–S. They make up 76% of the total contribution to λ. Adding the impact of a seed predator, *Rhinocyllus conicus*, to this system reduced the value of λ of this population from 2.2142 to 1.7078. Thus the population of the weed would continue to grow even with the biological control agent. This model does not incorporate density dependence and therefore any compensation that might buffer the impact of seed predators was not included.

Hal Caswell is the guru of matrix modeling and his book (Caswell 2000b) gives the most extensive coverage of this technique. Matrix models were also the focus of a special feature in *Ecology* (Vol. 81, March 2000) and can be explored using various software packages such as the RAMAS Ecolab (Schultz *et al.* 1999) and poptools (http://www.cse.csiro.au/client_serv/software/poptools/index.htm)

10 seeds to the density observed in center plots while 11 to 18 years were required for prairie plots.

To assess the potential impact of seed eating insects on introduced populations of broom Parker manipulated fecundity values and generated hypothetical λs for the slowest growing population and the fastest growing population. For the former 70% of the seeds had to be removed before λ became less than 1 and the latter populations would only decline when 99.9% of the seeds were removed. In this and other studies of broom in France (Paynter *et al.* 1998), Britain (Rees and Paynter 1997), and Australia (Downey and Smith 2000), disturbance of established populations seems to be the dominant factor in population dynamics. Survival of seedlings is largely dictated by whether mature plants are present in the area. However, Parker found that in prairies in western Washington the rates of increase of broom were greatest in undisturbed, open habitats. Therefore the dynamics of established and invading populations are influenced by different factors and these models were helpful in identifying these differences.

Combining population models and experiments

A model of biological control of *Sida acuta* in northern Australia

Combining experiments with models is an excellent approach to achieving a better understanding of invasive plants and their control. Several studies of invasive weeds have done this. For example, Lonsdale *et al.* (1995) studied a malvaceous tropical weed, *Sida acuta*, in northern Australia. A defoliating chrysomelid beetle, *Calligrapha pantherina*, was introduced as a biological control agent on this introduced weed and it reduced annual seed production from 8000 to 731 seeds m^{-2}. Beetle defoliation did not result in a measurable effect on individual plant survival, mass per seed, or total biomass of the weed in the year of defoliation. The model developed in this case was based on three equations.

$$N_{t+1} = \lambda N_t$$

where λ = (seeds/plant) (fraction viable seeds) (dry season seed survival) (fraction seeds germinating) (fraction seedlings surviving).

Growth is regulated through density-dependent fecundity.

$$S = s(1 + a N)^{-b}$$

where S = number of seeds produced per plant at a given density, s = number of seeds produced by an isolated plant, a = area required by a plant to produce s seeds, b describes the effectiveness of resource uptake from that area. An upper limit is placed on density by mortality due to self-thinning such that the density of flowering plants, N, and the initial density of plants, N_i, are described by

$$N = N_i(1 + mN_i)^{-1}$$

where m^{-1} is the asymptotic density of N as N_i tends to infinity. This leads to a combined equation of

$$N_{t+1} = \frac{\lambda N_t}{(1 + a N_t)^b + m \lambda N_t}$$

Parameters for the model were based on field measurements with the addition of an adjustment factor to bring the simulated population density to observed levels.

To determine the potential impact of the introduced beetle, plots were treated with insecticide. The impact of defoliation was measured by reduced seed production of plants in areas of beetle defoliation. Simulations using parameters arising from insecticide treated plots resulted in stable populations while those based on parameters from defoliated plants showed declining populations (Figure 8.6).

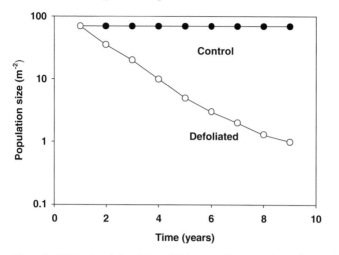

Figure 8.6. Simulated densities of *Sida acuta* for control conditions (beetles removed) and based on the reduction in plant fecundity following beetle defoliation. Closed circles represent control values and open circles incorporate defoliation. After Lonsdale *et al.* (1995).

By 1999 a review of the impact of *Calligrapha* on *Sida* found that in some sites *Sida* density was reduced by 84–99% (Flanagan *et al.* 2000). Beetles were most effective at coastal sites in northwestern Australia where the plants behave as annuals. Responses to questionnaires sent to land owners indicated that *Sida* infestations had been significantly reduced as had the costs for its control, but the success depends on when *Calligrapha* returns after the dry season. Unfortunately, data have not been published that could allow the model to be tested more thoroughly.

A model of biological control of tansy ragwort control in Oregon

The model above used the measured impact of a biological control agent on the fecundity of a plant and projected the longer term impact of this reduction on host plant density. Another approach is to take a successful biological control program and evaluate the various components that may have contributed to its success. This approach was taken by McEvoy and Coombs (1999) who studied the successful biological control of tansy ragwort, *Senecio jacobaea*, in Oregon. In this study they combined experimental manipulations and the use of matrix models to evaluate how disturbance, colonization, plant competition, and herbivory by two introduced control agents, cinnabar moth, *Tyria jacobaeae*, and flea beetles,

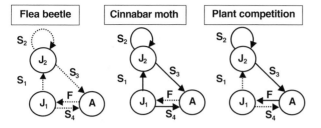

Figure 8.7. Life cycle graphs for experimental populations of tansy ragwort, *Senecio jacobaea*, an introduced pasture weed in Oregon. Stage classes in circles are $J_1 =$ 1-year-old juveniles, $J_2 =$ 2-year-old juveniles, and A = adults, S_1–S_4 are survival probabilities and F is stage-specific fecundity. Dashed arrows indicate transitions reduced by the treatment. Attack by flea beetles, *Longitarsus jacobaeae*, reduced all transitions among plant stages. These have successfully controlled ragwort in many locations. Cinnabar moth only reduces plant fecundity. Competition from other plants reduces ragwort growth. After McEvoy and Coombs (1999).

Longitarsus jacobaeae, influence the population growth of the introduced, target weed. Perturbation analysis (Caswell 2000a) was used to first assess how changes in the vital rates would influence changes in population growth that might occur, and then to assess the changes in populations that actually did occur due to treatments. These evaluations were based on the results from 96 experimental plots in which plant competition was modified by clipping and levels of attack by the two biological control agents were modified by caging (McEvoy *et al.* 1993).

McEvoy and Coombs (1999) used life table response experiments (Tuljapurkar and Caswell 1997) to measure the vital rates in experimental treatments and matrix models to project the changes in these rates on population growth. This procedure identifies the life-cycle transitions and paths that contribute the most to the growth of the weed populations. The impacts of the two biological control agents and plant competition on the vital rates of tansy ragwort life stages are diagramed in Figure 8.7. Flea beetles had a negative impact on all transitions while cinnabar moth only reduced fecundity. A conclusion from this analysis is that by considering the top-down (herbivore limitation) and bottom–up (resource limitation) (see Chapter 5, Figure 5.1), the number of control agents required to reduce the weed can be minimized. This result was also seen in the original experimental analysis (McEvoy *et al.* 1993) and is confirmed with the matrix models.

Given that each introduction of a new biological control agent has risks of non-target impacts associated with it, an experimental analysis of

Box 8.2 · *Determining mean values of* λ.

$$\text{Geometric mean} = \left(\prod_{i=1}^{n} x_i \right)^{1/n}$$

$$\text{where as the arithmetic mean} = \left(\sum_{i=1}^{n} x_i \right) / n$$

In projecting future populations, $N_t = N_0 \lambda$, usually an average value of λ is used as determined from survival and fecundity parameters. In many models λ is estimated from measurements in one year or by taking an arithmetic mean based on populations in several locations. However, because λ will be likely to vary over time for environmental as well as biological reasons, the projection based on an arithmetic mean of the population growth rate will not correctly project population numbers. The geometric mean will be lower than the arithmetic mean.

the potential impacts of species of agents is very useful. This study should be a model for future work.

The world is variable but models are not

Population models often ignore temporal variability in life history parameters. If more than one estimate of population growth rates exists, they generally use the arithmetic mean of these rates to predict future population abundance. Freckleton and Watkinson (1998) show that for stochastically varying environments, the persistence of weed populations depends on the geometric mean of population growth being greater than zero rather than the arithmetic mean being greater than zero (Box 8.2). Models based on the arithmetic mean of population growth rates will overestimate population size. Simulations by Freckleton and Watkinson show that those biases can be considerable. The largest effects of temporal variability on population estimates occur at low density. Long-term studies are necessary to estimate the variance of population growth rates.

Modeling invasive plants – what have we learned?

Models will always be limited as to how much of the stochastic aspect of nature they can incorporate and in how many of the biological interactions they can accommodate. A useful model will be one that is

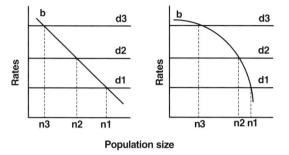

Figure 8.8. Change in equilibrium population density resulting from density-independent mortality (d) and linear or curvilinear density-dependent fecundity (b). After Watkinson (1985).

designed to integrate information to answer a particular question such as 'does reduction in seed production reduce plant population density?' or 'will increased competition enhance the impact of a control agent on reducing plant density?' In addition to the models, experiments and long-term density data are imperative for interpreting and testing predictions. A successful model is one that makes testable predictions. It is perhaps unrealistic to expect models to make accurate quantitative predictions, but it would be satisfying if more models were tested with field experiments or observations.

The response of plant survival and fecundity to varying density is something that should be kept in mind when developing models of plant populations. Not only is it important whether survival and fecundity are related to density, but the shapes of these relationships, whether they are linear or curvilinear, are important for predicting the fate of populations. The difference in population sizes with linear or curvilinear birth rates is diagramed in Figure 8.8 (Watkinson 1985, Lonsdale 1996). Birth rates and death rates should be obtained experimentally and incorporated into the models to take into account the potential for compensation in biological systems. For example, most models find that seed destruction would have to exceed 95% to reduce population density. These estimates might change, however, if the relationship between density and seedling survival is non-linear.

In an ideal world, field data would be collected before model development, models would make testable predictions and field data would be collected to test these predictions. Even in the real world, the process of modeling a population forces one to think about the important interactions in the biological system. For that reason alone models are extremely valuable.

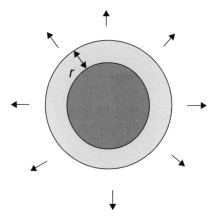

Figure 8.9. The simplest form of spread of an invasive plant is diffusion from a single focus in the middle of suitable habitat. In this diagram *r* measures the advancing front each year.

Modeling invasions as they spread across habitats and landscapes

Quantifying patterns of spread of a newly established species is a prerequisite to the development of effective procedures for both *delaying* the rate of spread of an introduced species and *predicting* its behaviour through time. Cousens and Mortimer (1995) point out that a lack of quantitative data hampers the precise calculations of the rates of spread of weed species. Data that do exist show that actual patterns of spread are complex (see Figure 5.5, showing the spread of knapweed along a hillside). In addition, other researchers also point to the need for a better understanding of how biological invasions proceed (Ehler 1998, Fagan *et al.* 2002). Daehler (2001) suggested that an effective way of identifying introduced species of concern is to identify those that are spreading rapidly. New technologies such as GPS and GIS have the potential for rapidly mapping the spread of species (see figures in Chapter 2 on Japanese knotweed and water milfoil). However, this kind of mapping is 'after the fact'. It would be more useful if mapping could be incorporated into an early warning system that could allow managers to deal with invaders in the early stages of establishment.

The simplest way of envisioning a spreading species is to imagine it spreading outwards along a front, at a constant rate in all directions (Figure 8.9). This process can be modeled as a diffusion process (Skellam 1951, summarized in Cousens and Mortimer, 1995).

If the distance moved each year by the advancing front is r, and we assume that the spread starts from a single point (figure above), then the area A occupied after t years will be:

$$A = \pi (r\,t)^2$$

for which the rate of increase in the area is

$$\mathrm{d}A/\mathrm{d}t = 2\,\pi r^2 t$$

and the instantaneous proportional rate of increase is

$$(\mathrm{d}A/\mathrm{d}t)/A = 2/t$$

Thus, the percentage increase of the area occupied will decrease over time.

However, most invasions do not proceed along a single radiating wave front, as modeled by Skellam (1951) (Moody and Mack 1988, Cousens and Mortimer 1995). There are usually multiple introductions in multiple locations (Novak and Mack 2001). Long-distance dispersal events, even when they are rare relative to the local spread of propagules, may effectively determine the rates at which a species moves across a landscape (Kot et al. 1996, Buchan and Padilla 1999, Fagan et al. 2002). The role of humans, in aiding transport, may be critical for increasing the rate and scale of invasions (Chapter 2 and Mack et al. 2000, Fagan et al. 2002). In fact, when long-distance dispersal is sufficiently common, the wave speed of a modeled invading population accelerates over time (Fagan et al. 2002). Invasions are, thus, increasingly being viewed as a metastatic process involving jumps to new sites, some of which might be quite distant from the original location.

The concept of 'nascent' foci

In an effort to shed some light on how weed control and eradication programs could be made more effective, and to take a more realistic approach to the dynamics of spread, Moody and Mack (1988) took account of the fact that there are usually multiple patches of weeds, rather than one discrete population (Figure 8.10). Weed control programs usually focus on the most conspicuous patches of invaders, and large patches or 'foci' are often the main, and in some cases the only, targets (Moody and Mack 1988). In order to evaluate the significance of these smaller patches of weeds, that they call 'nascent foci', they extended the models of Auld et al. (1978–79) and investigated the rates of spread of an invader from multiple, initially small foci, in a series of simulations.

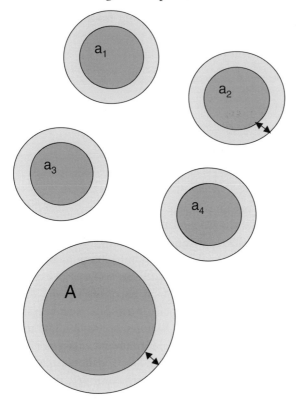

Figure 8.10. Spread from 'nascent foci' requires consideration of r, the annual spread, in small patches as well as large patches.

Imagine one patch of a weed, with a particular area, A, and imagine having four smaller patches, which, when their areas are summed, give a total area the same as the one large patch: $(A = a_1 + a_2 + a_3 + a_4)$ (see Figure 8.10).

The area of the one large patch is $A = \pi D^2$, while the area of each of the smaller patches is $a = \pi d^2$ in which D and d are the radii of the large and small circles.

It follows that $\pi D^2 = 4\pi d^2$, and that $D = 2d$. After simple diffusion of a distance r we can see that one time-unit later, the single source will cover an area $\pi(D + r)^2$.

However, each small source will individually have expanded to cover an area of $\pi(d + r)^2$.

The sum of the four smaller areas is

$$4\pi(d+r)^2$$
$$= 4\pi(D/2+r)^2 \text{ which can be reduced to } \pi(D+2r)^2.$$

Comparing the total area covered by the four small patches of the weed, with the increased area after one year covered by the one large patch, it is clear that the total area of the four small patches is greater, when no overlap of patches is assumed:

$$\pi(D+2r)^2 > \pi(D+r)^2 \text{ (Auld } et \text{ al. 1978–79, Mack 1985 reviewed}$$
in Cousens and Mortimer 1995).

It is clear from this simple model that, over the same time period, many small patches will spread more rapidly to cover a total area that is larger than that covered by one large patch, even when the total area of the large patch starts out being the same as the areas of all the small patches summed.

Moody and Mack (1988) extended the model to different scenarios, where small weed patches were either targeted for control or ignored:

1 Only the large patch was targeted for control, with the removal of the outer ring or patch edge.
2 Smaller patches were targeted for removal and the large patch was ignored.

If there is no control at all, then the area covered by the invading weed after some time unit, t, would be:

$$A = \pi(D+rt)^2 + n\pi(d+rt)^2$$

where $n =$ the number of small satellite patches.

If there is targeted control of only the large patch, the total area covered after time, t, is greater than if the small patches are controlled (Moody and Mack 1988). However, they noted that small, nascent foci or patches can re-establish from the 'mother' patch, so even if they are the target for eradication, repeated control is necessary. They concluded that small patches cannot be ignored, and that it is most efficient to detect and eradicate small patches at a *subcritical* stage. Interestingly, they point out that added factors such as the presence of geographical corridors which might facilitate spread or the inclusion of barriers, increased the mathematical complexity of the model without a corresponding increase in information (Okubo 1980 cited in Moody and Mack 1988).

Humans have become the major vector of species introductions for the past 200–500 years (see Chapter 2 and Mack *et al.* 2000), and through this 'active' transport clear 'nodes' of origin can be identified. The importance of long-distance dispersal events was evaluated by Buchan and Padilla (1999). They compared the predicted rates of spread for zebra mussels (*Dreissena polymorpha*), based on a simple reaction-diffusion model, with actual distances traveled by boaters who could be transporting the mussels as they moved their boats between lakes. The simple models *underestimated* the maximum spread and geographical extension of the range of zebra mussels. Depending on the diffusion coefficient used, the model predicted invasion velocities varying from 1.0 km yr^{-1} to 3.8 km yr^{-1}, based on surveys of distances traveled by boaters between lakes. This analysis generated shorter distances than the maximum distances over which zebra mussels were actually transported to previously uncolonized lakes. However, the model also *overestimated* the invasion of suitable habitats within the predicted geographical range of the zebra mussels. The model predicted that 35 boater-accessible lakes would become colonized by zebra mussels by 1997 and this was not the case. Their research highlights the importance of rare, long-distance dispersal events.

Theoretical models of the spread of introduced organisms have also explored the potential impacts on populations at low density of having reduced rates of population growth when an Allee effect sets a minimum population size for spread (Lewis and Kareiva 1993). The strength of this effect might be expected to vary between plants that are self-compatible, outcrossing or for which vegetative reproduction is important. Also, Lewis (2000) considers the role of intrinsic stochastic factors as might arise from the interactions among individuals in small local populations. His analysis shows that, compared to the results of a purely deterministic model, density-dependent interactions in high density, widely spaced foci could slow the spread of invaders. This would be the case in which plants growing at high density had reduced seed production and therefore potentially reduced spread. However, if the spread is limited to patches of suitable habitat, these interactions within patches may have little impact on the chance for long-distance movement of seeds. For example, seeds being carried on machinery or grazing animals to new foci could be more important to the spread than localized seed production and seed rain.

Theoretical models of the invasion process are limited only by the imagination while their application is limited by the ability to measure relevant parameters in the field. Apart from the lack of accurate data on the rate of spread of invasive plant species, traditional approaches for

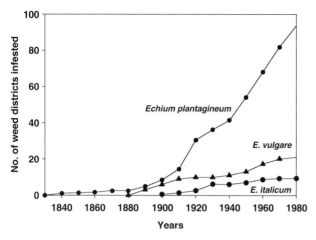

Figure 8.11. The spread of three species of *Echium* in Australia measured by the number of districts (1^0 latitude $\times$ 1.5^0 longitude) infested. This shows a long period of low density before expansion for *E. plantagineum,* but continued slow spread for the other two species. After Forcella and Harvey (1983).

modeling the spread of introduced plant species may not be appropriate (see Higgins *et al.* 1996, Kot *et al.* 1996). The real test of how useful models are is whether they generate testable predictions or tell us what to do.

What models tell us about detecting invasions

There is general agreement that detecting incipient invasions remains a severe challenge (Cousens and Mortimer 1995, Fagan *et al.* 2002). Usually, the species is present at low densities, before its rapid expansion is detected (Figure 8.11), and people only notice the presence of larger, more visible patches (Moody and Mack 1988). Fagan *et al.* (2002) point out that invasion theory is likely to be of limited use in the development of better detection protocols, but it may be helpful in providing guidance on the nature of spreading populations. For example, a plant species that is becoming an invader may have an accelerating rate of spread, which may exceed a predetermined threshold for an acceptable rate of spread, thus generating a call to action. However, this presupposes that there are reliable data on movement. There is, however, general agreement that earlier interception of small patches of invasive species is likely to be the most effective way of controlling spread across the landscape. Quarantine procedures can reduce the opportunities for long-distance dispersal (see Chapter 2, Moody and Mack 1988, Buchan and Padilla 1999).

Identifying sites of introduction is best done at ports of entry, and in corridors for dispersal (OTA 1993). The pattern of spread will subsequently be determined by whether the dispersal is influenced by wind, by animals, by contamination of agricultural products, by movement through water, or with movement of vehicles such as trains, construction equipment, logging equipment or boats. Buchan and Padilla (1999) point out that where the movements of people are likely to be the chief vectors of spread, as with zebra mussels, monitoring human movement, rather than looking for their occurrence in some previously uncolonized inland lake, may actually be the best way of predicting long-range dispersal. In another study, Buchan and Padilla (2000) used human activity to predict in which lakes a different aquatic invader, water milfoil, would next appear. They built a logistic regression model to look at the predictive power of three different kinds of variables:

1 those affecting human lake access, e.g. distance from highway,
2 those influencing milfoil growth conditions, e.g. pH, phosphorus,
3 those that attract human activity to a lake, e.g. presence of game fish.

When they compared the model with results of a lake survey, they found that variables associated with water quality were actually the best predictors of finding milfoil in any given lake.

A common pattern is for invasive species to spread from a site where the species has been planted such as a forest tree in a commercial plantation. This pattern of spread may be simpler to model than one involving patches of suitable habitat. In South Africa pine trees (*Pinus* sp.) have been spreading into the native Fynbos biome. Higgins *et al.* (1996) developed a spatially explicit, individual based simulation model (SEIBS) of this system. Their aims were to (1) compare the results of this model to a standard diffusion model, (2) explore the interactions among the five factors of adult fecundity, dispersal distribution, age of reproductive maturity, fire return interval, and fire survival, and (3) investigate the sensitivity of the model's predictions to the various factors. The SEIBS model was a 2^5-factoral simulation in which they considered 32 different combinations of the five life history and fire related factors. It was based on a grid starting with 150 cells (trees along the edge of a plantation) along the y axis and 400 cells (movement spaces) along the x axis. For each combination they determined a finite rate of increase (λ) and a diffusivity (m^2 yr^{-1}) index based on the mean dispersal distance and the fire return interval (germination could only occur after fire) used in the simulation. For the diffusion model the asymptotic rate of spread, $V = \sqrt{4rD}$ where

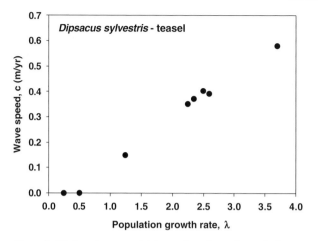

Figure 8.12. Invasion speed (c) as a function of population growth rate λ for eight populations of teasel. After Neubert and Caswell (2000).

$r = \ln(\lambda)$ and D = diffusivity. They concluded from this comparison that reaction-diffusion models are inadequate for modeling spatial phenomena. The SEIBS model, however, was able to mimic ecological processes and interactions. The results of the model were sensitive to the spatial grain and the authors recommend that the spatial grain of the model must represent the patterns of environmental heterogeneity. The most important factor in the model was dispersal ability and it would seem that this is exactly what one would want to know. The authors conclude that parameter estimation and model development processes must be integrated. However, it remains a large question if these types of models will ever be able to tell us things that we did not already know in relation to the spread of an alien species. The ability to disperse is the metric of invasion. This must be measured and will not arise readily as a result of a simulation model.

Invasion speed for structured populations

Population growth and dispersal determine the speed of plant invasions. Neubert and Caswell (2000) constructed a discrete-time model that couples matrix population models for population growth and integrodifference equations for dispersal. They apply these models to data on two plant species, *Dipsacus sylvestris*, a perennial herbaceous plant of old fields in northeastern North America and *Calathea ovandensis*, an understory plant of successional, neotropical forests. Wave speed and population growth (λ) were correlated (Figure 8.12) as were their elasticities and sensitivities.

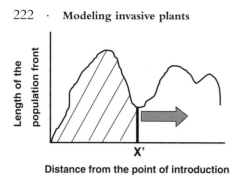

Length of the population front

X'

Distance from the point of introduction

Figure 8.13. The distribution of an introduced plant species will include areas of suitable habitat already occupied (diagonal lines) and those still unoccupied. The length of the population front will determine the costs of creating barriers to slow the spread of the species. Models can be developed to evaluate the costs and benefits of slowing the spread of an introduced plant species. After Sharov and Liebhold (1998).

In addition, when dispersal was made up of both long- and short-distance components, the long-distance components determined invasion speed even when these were rare. These models, like those described above, of course lack the complexity and stochasticity that are likely to influence the rate of spread of introduced plants. They do represent another approach to this important dimension of invasive plant species.

Slowing the spread

Evaluation of the economic benefits of slowing the spread of invasive species has been the focus of models developed by Sharov and Liebhold (1998). Although they were particularly concerned with the spread of exotic insects the same principles could be applied to plants. Pest introductions can occur in three situations: in the center of the potential species range, at the edge (e.g. along the coast, where species can only move in one direction), or at one end of a linear area. The length of the population front and the areas that are occupied and unoccupied (Figure 8.13) influence the cost and benefits of control measures. It is possible that in some areas the length of the population front will be constrained and this could reduce the costs of creating a barrier zone to slow the spread. These types of models force managers to consider the biology of the invasion, e.g. how the plant is spread, and the costs and benefits of attempted eradication or in creating and maintaining barriers.

Conclusions

In a treatise on theory and data related to population dynamics in spatially complex environments, Peter Kareiva (1990) makes the following statement:

The challenge for empiricists is to investigate more rigorously the roles of spatial subdivision and dispersal in natural communities. The challenge for theoreticians is to make the empiricist's job easier; this can best be done by delineating when spatial effects are most likely to be influential, and by offering guidance on how to design appropriate experiments. Simply saying that the spatial environment is important is to mouth a platitude: what we need to know is whether this presumed importance amounts to much in natural systems. (*p. 53*)

It is not clear that thus far the theoreticians have made the job of the empiricists easier but students are challenged to discover if this is so. The most useful models are those that are closely related to field situations. These can be used to make realistic projections. We know that the invasion of plants is not a simple diffusion process and that spread of plants will occur by the establishment of foci followed by diffusion. There continues to be a lack of guidance from the theoreticians as to how the empiricists should study the spread of introduced species. One must be explicit about the questions to be asked and the predictions to be made. The answers are likely to be the result of hard fieldwork. The critical practical questions about invasion will be things like how rapidly is a species spreading? Do control measures reduce the rate of spread? Can species invade undisturbed as well as disturbed sites? Does density related reduction in seed production reduce the rate of spread? Models are often useful in telling us what we do not know or what doesn't work. Progress can be made best when empirical work is a major component of the modeling exercise and predictions are explicitly stated and tested.

9 · *Action against non-indigenous species*

Introduction

We have previously discussed the historical, socio-economic, and ecological aspects of introduced plants. However, land managers and environmental consultants are faced with developing plans and procedures for responding to invasive plants. For them conservation and restoration are the goals. Some researchers would, no doubt, argue that a clear economic cost–benefit analysis of different strategies is essential prior to embarking on a weed control program (e.g. McNeely 2000, Naylor 2000). However, even with such an analysis, success cannot be guaranteed (Rea and Storrs 1999). Science can help to inform management decisions, but ultimately all such decisions about our environment are driven by political imperatives, social concerns and economic constraints. Although the latter factors often dominate, management decisions should still be based on good ecological principles when these are well established and requires, at the very least: (1) sound, defensible science, and (2) education about the issues.

In this chapter we summarize the techniques and programs that have been used to deal with introduced species, and the processes by which habitat managers and researchers may evaluate these programs. We also look at how management strategies have been developed or are developing at international, national, regional, and local scales. In this chapter we broaden our outlook to include considerations relating to non-indigenous species that are not purely ecological. We consider the perceived benefits of introduced plant species, as well as the costs, benefits and outcomes of large-scale and small-scale removal programs for invasive species.

The scale of the problem

Too often a single problem species causes the call to arms to 'do something'. But, as outlined in previous chapters, invasive species do not occur in an environmental vacuum and restoration of pre-invasion conditions

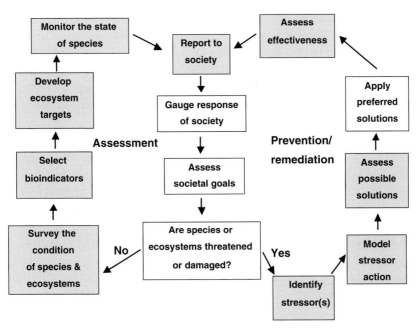

Figure 9.1. Representation of a model plan for ecosystem management. Shaded squares require input from ecologists. After N. Yan P. Dillon, G. Mierle, and K. Somers (personal communication).

will inevitably be complex and slow if even possible. Restoration may require decades to centuries (Lovich and Bainbridge 1999, McLachlan and Bazely 2001). Initiatives aimed at controlling non-indigenous species range from legislation, to strategic policies, to practical advice on how to get rid of Japanese knotweed, *Fallopia japonica* (Coleshaw 2001). Many approaches and strategies have a single-species focus. A similar approach dominates the conservation of rare and endangered species (but see Noss 1996). Single-species actions are sometimes inappropriate because the invaders are flourishing due to changes to ecosystem processes such as fire and hydrological regimes (Chapin *et al.* 2000). Additionally, some invasive species function as 'keystone' species and modify ecosystem functioning (Chapter 3). Thus, even their removal may not cause a return to the 'uninvaded' state. Therefore it makes sense to take an ecosystem approach to management and Noss (1996) has called for ecosystems *per se* to be conservation targets rather than individual species.

A model plan for ecosystem management has been developed by Norman Yan (personal communication) and provides a clear and effective approach to dealing with this issue (Figure 9.1). This procedure is built on continual assessment, evaluation, prevention, and remediation. There is a major role for ecologists in this process. Ecologists, managers, and the public must all be involved.

When restoration is attempted, of necessity there must be a vision in mind, and this is usually to return the damaged ecosystems to a condition that is structurally and functionally similar to that before the disturbance. Defining this pre-disturbance state can often be a challenge! European ecosystems have been modified by humans for millennia. For North American ecosystems, the state prior to European settlement is often defined as the target. However, the plant community composition before disturbance is generally not known for a particular site, and it certainly will not have been static. An excellent handbook for restoration ecology is the Society for Ecological Restoration's *Primer on Ecological Restoration* (SER 2002). This contains considerable useful information for managers.

Manuals and advice

A large number of internationally focused books outline the range of control methods that have been developed for different invasive plants (Cronk and Fuller 1995). Locally based manuals also explain what has been learned through trial and error about the best way to control a particular species (e.g. Bossard *et al.* 2000, Coleshaw 2001). English Nature in the United Kingdom has developed a 'Practical Solutions' Handbook describing the ecology of various problem plant species and offering advice on methods used to control them (Coleshaw 2001). A similar book for California, USA, explains the biology and control of the 78 non-indigenous plant species listed as being of greatest ecological concern in California by the California Exotic Pest Plant Council (Bossard *et al.* 2000). Thus, there is a wealth of information, based both on practical experience, and experimental research, from people who have tried to get rid of a problem plant.

Two general conclusions can be drawn from the literature on control of environmental non-indigenous species, as opposed to agricultural weeds:

1 Standard control techniques are generally both costly and labour intensive. Habitat managers who head down the control path must be aware of the size of the task ahead.
2 Integrated weed management (IWM) that may involve several types of techniques, is the path of choice (Goodall and Erasmus 1996, Dozier *et al.* 1998, Mullin *et al.* 2000). Biological control is generally viewed

as an essential part of a multi-faceted control strategy (Laroche 1998). A common refrain in many articles is the explicit wish for an effective biological control agent (e.g. Loope *et al.* 1988, Braithwaite *et al.* 1989, Dozier *et al.* 1998, Laroche 1998, Turner *et al.* 1998, Holmes *et al.* 2000) (see Chapter 7). However, other approaches to control are also feasible and even preferable in some cases. These are considered next.

Physical control methods

Physical controls include both targeted and non-targeted methods. In non-targeted methods, an ecosystem-level factor, such as disturbance, is manipulated to achieve the desired outcome of reducing cover and density of the target species. Targeted methods of physical removal include cutting and pulling a problem invader. Non-targeted methods include burning, grazing and flooding to alter the disturbance regime.

Pulling and cutting

The appeal of cutting and pulling programs is that they allow volunteers to get involved in habitat restoration. The timing of pulling and cutting is very important. If plants are pulled out while they are flowering, their transport and removal may spread seeds far and wide. For the invasive shrub Scotch broom, *Cytisus scoparius*, in British Columbia, pulling plants resulted in soil disturbance and trampling which facilitated recovery of plants from the seed bank (see Chapter 8 for a discussion of broom and disturbance). Cutting was clearly much more effective (Ussery and Krannitz 1998). Loope *et al.* (1988) list a number of species present in various arid land reserves in the USA which were being controlled by hand pulling. Extensive hand pulling has also been carried out at Point Pelee National Park, Canada (G. Mouland, personal communication). This is labour-intensive work and is really only feasible when cover of an invasive species is low. Comparisons of different cultural control methods for Japanese barberry, *Berberis thunbergii*, showed that pulling and cutting alone, were not as effective as cutting and applying a herbicide to stumps (Figure 9.2A). These examples serve to illustrate that it is always worthwhile carrying out some experiments to evaluate control methods.

Non-targeted physical control

While a number of invasive species are described as 'fire-enhancers', burning has also been recommended as a management tool for suppressing

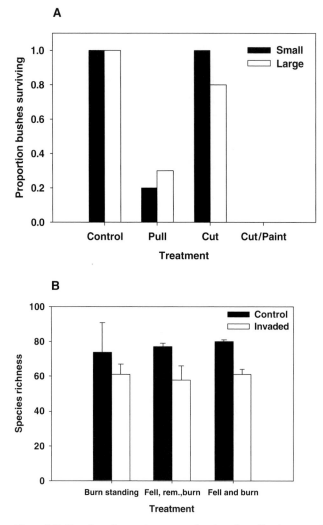

Figure 9.2. Results of experiments evaluating the effectiveness of different control techniques. (A) Targeted cultural control of large and small Japanese barberry, *Berberis thunbergii*, in Rondeau Provincial Park, Ontario. Treatments are to pull out the plants, cut them or cut and treat with herbicide (D. R. Bazely, unpublished data). (B) Effects of different types of burning management on the species diversity of fynbos vegetation in control, non-invaded, and invaded areas. Treatments are to burn standing vegetation, fell, remove and burn, and fell and burn. Invasion reduced biodiversity but burn treatments did not have a significant effect (Holmes *et al.* 2000).

non-indigenous species such as garlic mustard, *Alliaria petiolata* (Nuzzo 1991). Prescribed burning schedules can vary according to season, fuel load, and frequency. Many plant communities in North America, Europe and Australia are burn-dependent. As discussed in Chapter 3, altering disturbance regimes can have unexpected results. It is important to investigate variations on the same basic type of management, since they can have subtly different effects. South African fynbos vegetation is fire prone and fire adapted and in many areas contains introduced *Pinus* species. Three burning treatments were tested in this habitat: 'burn standing', 'fell and burn', or 'fell, remove and burn'. Treatments did not vary significantly in influencing species diversity or canopy cover (Figure 9.2B). However, the 'burn standing' treatment had less of an impact on other ecosystem variables than did the 'fell and burn', probably because the latter treatment caused a hotter fire (Holmes *et al.* 2000). The most important outcome of this research was the failure of burning to have the positive impact on native species diversity that was predicted. This indicated to managers that invasions of fynbos by introduced woody vegetation should be actively prevented.

It is vital to carry out experiments and monitoring to evaluate the outcome of burning as a control tool. Goodall and Erasmus (1996) recommend burning to control Siam weed (triffid weed), *Chromolaena odorata*, but Norgrove *et al.* (2000) found that burning did not control this species in their sites, as had been widely assumed by earlier workers. Similarly, Gentle and Duggin (1997) do not recommend burning to control *Lantana camara*. While fire seems to promote the conversion of sawgrass wetlands to a paperbark (*Melaleuca quinquenervia*) forest, burning did kill some small *Melaleuca* saplings (Turner *et al.* 1998).

The spread of *Melaleuca* followed large-scale alterations of the hydrological regime in the Everglades, Florida, USA, by the US Army Corps of Engineers. Flooding is being investigated as a means of control of this, and other species, and research is being conducted into the survival of *M. quinquenervia* seedlings under different hydroperiods (Lockhart *et al.* 1999a, Kaufman and Smouse 2001). David (1999) found however, that reintroducing natural hydroperiods did reduce *Melaleuca* but also eliminated bahia grass, *Paspalum notatum*, a species introduced for cattle forage, while torpedo grass, *Panicum repens*, another introduced species, formed dense monotypic stands in response to increased hydroperiod. Thus, management substituted one invasive species for another.

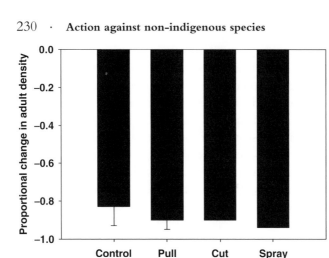

Figure 9.3. The effect of different control methods on the change in the density of flowering garlic mustard (*Alliaria petiolata*) stems from 1994 to 1995 in Rondeau Provincial Park, Ontario (mean ± SE, n = 9–10 plots).

In conclusion consistent and continued pulling can work to remove invasive species from small areas and even achieve very local eradication. But as long as there is a source of seeds or cuttings for reinvasion the control effort must continue. Larger community level perturbations require preliminary experiments to evaluate their likely success and possible problems. Experimentation and quantification through monitoring – adaptive management (Walters 1986) – provide the best approach.

Chemical control of non-indigenous plant species

A number of issues are associated with the use of chemical control. First is the possibility that non-target organisms will be affected (e.g. Rea and Storrs 1999). Additional problems are the need to have an approved herbicide (e.g. Goodall and Erasmus 1996), and the necessity to avoid the evolution of resistance. Chemical control is commonly recommended in conjunction with cutting (e.g. Goodall and Erasmus 1996) (Figure 9.2A). Widespread aerial spraying is not often used, but it has been carried out for *Melaleuca quinquenervia* (Laroche 1998). Spray drift can be a serious problem for aerial weed control in agricultural areas.

An interesting example of why all control methods should be field-tested is given in Figure 9.3. Regardless of whether garlic mustard plants were pulled, cut to ground level or sprayed with herbicide, the flowering plant density declined the next year by the same amount in all plots, including the untreated control plots. This was subsequently explained by the population dynamics of this species in southern Ontario, in which density fluctuates in alternating high and low density years (D. R. Bazely unpublished data).

Costs and benefits of control

Cost–benefit ratio analyses are rarely done before beginning a control or eradication program for environmental weeds. As a result the real monetary costs often come as a shock to those in the field attempting to control an invader. In agriculture, the overall aim of weed control is to minimize the impact of the weed on the value of the crop, rather than for their total removal or widespread reduction (Westbrooks 1991). Additionally, weed control in high intensity arable crops is likely to be less expensive than the costs of controlling environmental weeds when calculated on an areal basis, because effective treatment can take place both before and after the crop is grown. In natural habitats, weed control techniques developed for agricultural fields, such as aerial spraying are generally not desirable, unless there are monospecific stands of the species, as is the case with *Melaleuca* (Laroche 1998).

Although the cost–benefit accounting done for intensive agricultural systems is much more difficult for natural systems (Naylor 2000), it has been attempted for *Tamarix* (Zavaleta 2000a,b and Box 9.1). In addition, as outlined in Chapter 7 cost–benefit calculations have been used for biological control programs, and they are positive, i.e. benefits > costs, for successful programs.

More often than not, managers of protected areas see that a plant is spreading and take an immediate decision to do something about it (Loope *et al.* 1988, Dunster 1989). Even more often, the public demands that something be done when they see waterways clogged with weeds or rangeland grasses replaced by poisonous weeds. However, evaluating success and effectiveness is essential and this should be planned at the outset. Frequently this evaluation is inadequate (Finlayson and Mitchell 1999, Rea and Storrs 1999). The results from such studies are important and revealing, and sometimes unexpected. In the sections below, research results from the control literature and recommendations in manuals are discussed in terms of assessment and monitoring of the effectiveness of control.

Box 9.1 · *Estimating the economic cost of* Tamarix *in the USA.*

The aim of Erica Zavaleta's (2000a) accounting exercise was to do a 'full-cost' accounting for *Tamarix* and to include the value of 'ecosystem services'. Ecosystem services are the conditions and processes through which species in natural ecosystems, sustain and fulfill human life (Daily 1997). Figure 9.4, adapted from Costanza (2001), illustrates a conventional economic view that does not consider ecosystem services and an ecological economics model for which community and individual well being are the end products rather than economic gain. Zavaleta explicitly sought to take an ecological economics approach and for this a number of techniques were used for generating different economic values, as outlined below.

Ecosystem service	Valuation method
Water provision	
Municipal value	Replacement cost
Agricultural value	Farm budget residual method
Hydropower value	Replacement cost
River recreation value	Contingent valuation/willingness to pay
Flood control	Avoided damages
Wildlife	Contingent valuation/willingness to pay

Zavaleta (2000a) assumed a discount rate of 0%. The cost of controlling *Tamarix* was estimated, in 1998US$, for a 20-year program that involved: site evaluation, eradication, revegetation, and monitoring. These costs were compared with the net economic benefits from *Tamarix* eradication calculated over a 55-year period. Some of the benefits included the projection that a number of planned water projects would consequently not be needed.

The continued presence of *Tamarix* was estimated to cost US$7–16 billion in lost ecosystem function over a 55-year period. The total net benefits of *Tamarix* eradication were estimated between US$3.8 and 11.2 billion, depending on how ecosystem values

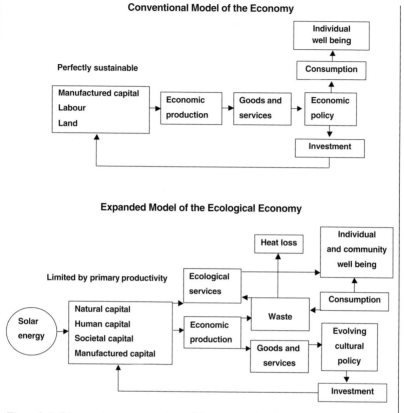

Conventional Model of the Economy

Expanded Model of the Ecological Economy

Figure 9.4. Diagramic representation of the conventional economic model and an ecological economic system in which limitations set by solar energy and waste production costs are taken into consideration. Greater emphasis is placed on community well being and evolving cultural policy rather than the more narrowly defined economic policy. Adapted from Costanza (2001).

were estimated. The benefit–cost ratio was in the range of 2.1–3.3. For a comprehensive explanation of these values, see Zavaleta (2000a,b).

Assessing control of non-indigenous species

Three important questions that should be asked about control strategies and programs are: (1) Have they been successful? (2) How was success measured? and (3) Can an invasive plant species be eradicated?

'Money is everything' is an apt expression to use when describing the likelihood of success of a control program. It is widely recognized in the literature, that control is likely to be successful only with enormous investments of time and money (Goodall and Erasmus 1996, Laroche 1998, Mack *et al.* 2000). Generally, success when it is monitored, is measured as a reduction in area covered by an invasive species. Programs do not explicitly aim for eradication, but rather 'maintenance control', which is often considered the only feasible option (Goodall and Erasmus 1996, Laroche 1998, Mack *et al.* 2000). However, in the long run eradication, if it is feasible, may actually be cheaper than control in the case of fast-spreading species (Cousens and Mortimer 1995, Box 9.2).

Given enough money, control can be highly effective. Francois Laroche (1998) reported that with an $11 million expenditure over a 7-year period, *Melaleuca quinquenervia* has been cleared from three water conservation areas in Florida, and is no longer increasing in the South Florida Water Management District. He projected that, if funding levels were maintained, *M. quinquenervia* could be eradicated from the Everglades Water Conservation Areas within 10 years. Similarly, intensive control programs for water hyacinth have reduced areal coverage from over 500 000 acres in Louisiana and 125 000 acres in Florida to a few thousand acres in those states (Mullin *et al.* 2000). Unfortunately, these reports are vague when describing the nature of the data and whether they are quantitative or qualitative. The monitoring protocols are also not often clearly articulated and Finlayson and Mitchell (1999) argue that little quantitative monitoring is done.

Eradication as a goal

Is total eradication possible? To our knowledge, no non-indigenous plant species that became invasive and covered large areas has been successfully eradicated (e.g. Mullahey *et al.* 1998) (see Box 9.2). Wild red rice may be the one confirmed example of a successful plant eradication (Mack *et al.* 2000). However, this took place *before* the species had spread widely (Table 9.1). This species was not initially identified (Vandiver *et al.* 1992), but its slow spread facilitated its removal and may indicate that it was not well adapted to the area. Thus, control and removal of small populations is more likely to be successful than removal of large areas of plant cover (Moody and Mack 1988, Hall *et al.* 1998).

Box 9.2 · *Eradication.*

Frequently programs initiated against introduced plants are referred to as 'eradication' programs. The definition of eradication may vary but that proposed by Myers *et al.* (2000) is the 'removal of all individuals of a species from an area to which reintroduction will not occur'. Eradication must be distinguished from population suppression or control which is the reduction of the density of the target weed or its rate of spread. The goal of a project may be severe reduction, but claiming eradication as the ultimate goal is optimistic for any species that has spread beyond a very restricted area.

In a recent review of eradication (Myers *et al.* 2000) proposed that six factors were necessary for a successful eradication program:

1 Sufficient resources to complete the project;
2 Clear lines of authority for making decisions on program operations;
3 A target organism for which the biological characteristics are *compatible* with eradication (easy to find and kill, little or no seed bank);
4 Easy and effective means to prevent reinvasion;
5 Easy detection of plants at low densities; and
6 Plans for restoration management if the species has become dominant in the community. There is little value in replacing one weed with another.

New Zealand has developed a plan for dealing with introduced species which is a model that could be of value to other jurisdictions (Myers and Hosking 2002). This plan consists of the following stages:

Detection – By having a program in place to detect expected invaders action can be initiated before a species spreads.

Evaluation – This should include delimitation of the distribution of the species, site description, impact assessment, response option identification, consultation, and the identification of the project leadership team. Consultation must be open and inclusive with input from industry sector groups, specialists, government departments, and the public.

Response Decision – The response decision group (recommended to be no more than 12 people representing science, policy, and operations) will consider the options arising from the technical and sector input in the evaluation report.

Operation Phase – This will include a statement of clear objectives, a communication plan, an operating plan, a consideration of health and environmental implications of the program, clear documentation of decisions and procedures, and an outline for financial management including an audit.

Monitoring Phase – Monitoring the course of the operation and its accomplishments at completion is imperative. This will include changes in the target pest population as well as impacts on the environment, human health, and community support.

Review Phase – To learn from eradication programs it is essential that information is made readily available through the formal publication of results. The review phase should be a clearly budgeted item. Failures and successes should be documented and suggestions for change made.

Area-wide management and slowing the spread

For widely established weeds, area–wide management may be a more realistic goal. This will involve continuous efforts to suppress populations of the plants chemically, mechanically and/or biologically in all locations. Working along the borders of the plant distribution may help to reduce the spread of the invasive weed, but metastatic spread of an invasive exotic following seed dispersal by animals or movement of plants by human activities works against successful containment (Chapter 8).

In conclusion, rapid action following the identification of a new exotic may allow successful eradication, but procrastination can cause the window of opportunity to slam shut. Having a plan in place, including identifying a chain of authority and developing a list of anticipated invading species, is effective in dealing rapidly with new introductions. In general eradication, even in localized areas, is not a feasible goal for dealing with invasive plants since reinvasion is likely to occur or not every individual including seeds in the seed bank can be removed.

These scanty data beg the question of whether eradication should ever be a management goal. Again, given adequate resources, it could be possible in some situations. Interestingly, even as Westbrooks (1991) states that widespread weeds cannot be eradicated, he calls for eradication of some species (see Tables 9.2 and 9.3). In the USA, 10 Federal noxious weeds are the targets for eradication from *localized* sites (Westbrooks *et al.* 1997). These are all cooperative projects between the federal government

Table 9.1. *Plant eradications reported in the literature*

Species	Location	Area	Eradication confirmed?	Source
Wild red rice, *Oryza rufipogon* (*Oryza sativa?*)	Everglades National Park, USA	0.6 ha	Yes – more than one non-review refereed source found	Vandiver *et al.* 1992 Westbrooks *et al.* 1997 Hall *et al.* 1998
Japanese dodder, *Cuscuta japonica*	Clemson, South Carolina, USA	1 ha	No – more than one non-review refereed source could not be found	Westbrooks *et al.* 1997 cited in Mack *et al.* 2000
Witchweed, *Striga asiatica*	North and South Carolina, USA	Statewide	No – currently infestation is reduced to 6600 acres from 175 000 ha in the 1950s	Westbrooks *et al.* 1997
Fountain grass, *Pennisetum setaceum*	La Palma, Canary Islands, Spain		No – project began 1998, projected to be successful in 6 years	Anon. 1999

and the affected states (Table 9.2), and it is noteworthy that even in the richest country in the world, the emphasis is on local removal. The use of the term 'eradication' for removal of a plant at a local or regional level, rather than continental or island level, is confusing since, while it may be removed in one state or county, if it is present in another, there is always the chance of re-establishment (Box 9.2).

Increasing the chances of successful control

Non-indigenous species are now on the agenda of many different groups or 'stakeholders'. Strategic and hands-on policies are being produced to deal with invasive species. An enormous amount of university-based research is being done (reviewed in Lozon and MacIsaac 1997 and Binggeli *et al.* 1998). The agriculture and horticulture industries have taken note of the problem, as have botanical gardens (White and Schwarz 1998, Hitchmough and Woudstra 1999, Mullin *et al.* 2000). Invasive species

Table 9.2. *Noxious weeds in the USA that have been targeted by the US Department of Agriculture for eradication*

Species	Status
Witchweed, *Striga asiatica*	177 000 ha infested in NC and SC; now reduced to 11 000 ha in 17 counties in NC, and in three counties in SC
Branched broomrape, *Orobanche ramosa*	283 ha infested in Karnes County, TX
Goat's rue, *Galega officinalis*	16 000 ha infested in Cache County, UT
Mediterranean saltwort, *Salsola vermiculata*	550 ha infested in San Luis Obispo County, CA
Hydrilla, *Hydrilla verticillata*	310 km of canals infested in the Imperial Irrigation District, Imperial Valley, CA; now 99% eradicated
Japanese dodder, *Cuscuta japonica*	1 ha infested in the SC Botantical Garden, Clemson, SC
Small broomrape, *Orobanche minor*	Spot infestations in Washington County, VA; Pickens, Abbeville, and Aiken Counties, SC; and in Baker County, GA
Catclaw mimosa, *Mimosa pigra* var. *pigra*	405 ha infested in Martin and Palm Beach Counties, FL
Asian common wild rice, *Oryza rufipogon*	A rhizomatous red rice; 0.5 ha infested in the Everglades National Park, FL
Wild sugarcane, *Saccharum spontaneum*	A rhizomatous wild sugarcane; 13 spot infestations along the southeastern shore of Lake Okeechobee in Martin County, FL, totalling less than 1 ha

From Westbrooks *et al.* (1997).

warrant an entire section in the Society for Ecological Restoration's *Primer on Ecological Restoration* (SER 2002). This all seems cause for hope.

In a seemingly good news story, Mullahey *et al.* (1998) described how the Tropical Soda Apple Task Force, established in Florida, in 1987, worked with the livestock industry to develop best management practices aimed at reducing the spread of the highly invasive soda apple, *Solanum viarum*, which spread widely and rapidly in Florida after its introduction in the early 1980s. The area occupied rose from 10 000 ha in 1990 to 500 000 ha in 1995, and it now occurs in adjacent states. Interestingly, the fruit are consumed by cattle, so the spread was directly associated with livestock movement (Mullahey *et al.* 1998). Soda apple caused a clear cost to Florida farmers, estimated at $11 million annually, which was associated

Table 9.3. *Steps in developing an effective strategy for dealing with non-indigenous plant species*

Westbrook's issue	Possible action
Identify foreign weeds that should be excluded	Create international databases, e.g. Binggeli *et al.* 1998
Develop and implement effective methods to exclude foreign weeds	Adopt a risk assessment and screening process
Identify incipient infestations of new weeds	Create effective modes of communications for people in the field who will come across these plants
Need for resources to address 'new weeds', through research and regulatory initiatives	Fund monitoring (Finlayson and Mitchell 1999)
Develop criteria for determining which new weeds can and should be eradicated	
Determine who should pay for control costs and costs of eradication	

Adapted from Westbrooks (1991).

with lowered livestock carrying capacity and overheating of cattle due to their avoidance of shady hammocks infested by soda apple. Mullahey *et al.* (1998) described the situation as a 'biological and ecological nightmare' that could have been avoided if adequate risk assessment had been carried out prior to 1990.

There is no shortage of advice on how to proceed with preventing the introduction of invasive species. One such example is Westbrooks' (1991) list of six issues that need to be addressed if further introductions of invasive species into the USA are to be prevented, and, what he terms 'eradications' are to be carried out (Table 9.3). The problem, however, is that carrying out the advice is often beyond the financial means of the agencies or groups that wish to act.

Who should take responsibility for introduced species?

It is widely recognized that most non-indigenous *plants* are not invasive (Lockhart *et al.* 1999b). The problem is that usually the invasiveness of those species that do become aggressive weeds was not predicted or

Figure 9.5. Distribution of carrotwood, *Cupaniopsis anacardioides*, in Florida, USA (hatched area), and locations of nurseries growing the species (circles). From Lockhart *et al*. (1999b).

detected (Mullahey *et al*. 1998). We discussed in some detail in Chapter 4 attempts to develop systems to predict the species that will become invasive. Accurately predicting all invasive species is currently not possible. However, we have learned three clear lessons from past introductions:

1. The worst weeds were deliberately introduced and then escaped (Binggeli *et al*. 1998). Nearly all introductions of woody plants that have become invasive, were introduced by horticulturalists, botanists, foresters, agroforesters or gardeners. Figure 9.5 shows the spatial relationship between nurseries growing the carrotwood tree, *Cupaniopsis anacardioides* and locations where it has escaped into natural habitats in Florida, and has been found to be reproducing, or naturalized (Lockhart *et al*. 1999b).

Clearly, one useful direction is to charge the people and industries who are actively importing and introducing new species with the responsibility for any negative impacts. This will require new legislation and regulations that will be resisted by well-funded lobbyists. One paradox that needs to be explicitly addressed is that many of the attributes of invasive species are precisely those characteristics favored by the horticulture industry (White and Schwarz 1998).

2. Many invasive species have been repeatedly introduced without initial monitoring of the ecological consequences (Binggeli *et al.* 1998). Examples from the USA include paperbark tree, *Melaleuca quinquenervia* (Austin 1978), and cogongrass, *Imperata cylindrica* (Dozier *et al.* 1998). In 1936, Hully Sterling aerially broadcast *M. quinquenervia* seeds over large portions of the Everglades, Florida, from the air (Austin 1978). If a species is going to be widely spread, then it would make sense to do some initial testing of its impacts.

3. Alterations to natural disturbance regimes are frequently associated with the establishment and spread of introduced species (Lozon and MacIsaac 1997). DeRouw (1991) clearly showed that the main factor associated with the establishment of *Chromolaena odorata* was whether a field was cleared from primary or secondary forest. Records of these types of observations can be very useful to other managers.

A major issue in current discussions of guidelines for predicting the potential invasiveness of plants is the World Trade Organization agreements that prevent national and regional governments from introducing procedures that would restrict trade (see Chapter 4). Also there has been almost no discussion of who should take responsibility for the enormous costs associated with controlling those species that have been introduced intentionally and have then created environmental and agricultural problems. Perhaps the WTO should establish a fund for the purpose of controling escaped plants and thus compensating for the impacts of imported plants and diseases that have arisen through trade.

The uncertain status of some invasive species

Even as some scientists are developing strategies and control programs for invasive species and researching their impacts on plant communities and ecosystem functioning, other scientists are researching the benefits of introduced plant species. Most introduced species have not become invasive and, in fact, from a biodiversity perspective, introductions have enriched some floras such as that of central Europe (di Castri 1989). In addition, introduced species can have potential medicinal properties (e.g. Ghisalberti 2000, Ambika 2002) and agricultural benefits (e.g. Roder *et al.* 1997,

Norgrove *et al.* 2000). The ambiguity relating to how a species is valued is illustrated by the case of Siam weed or triffid weed, *Chromolaena odorata*, a widespread weed of the Old World tropics and a perennial, semi-woody shrub of the neotropics (Cronk and Fuller 1995). Dates for the introduction of *C. odorata* vary according to the source used. Binggeli *et al.* (1998) report its introduction to botanic gardens in India and Sri Lanka in the nineteenth century and to Southern Africa in the early twentieth century. A later round of introductions to southeast Asia and many other parts of the world, including west and south Africa, occurred in the 1930s and 1940s (Goodall and Erasmus 1996, Roder *et al.* 1997, Binggeli *et al.* 1998). *Chromolaena* is now widespread in areas used for slash and burn agriculture and in agroforestry ecosystems, where trees and understory crops are grown together (Roder *et al.* 1997, Norgrove *et al.* 2000). In Laos it colonizes upland rice fields during fallow periods, which have declined from 38 years prior to the 1950s to 5 years in 1992. Nevertheless, despite farmers considering weeds to be one of three main factors limiting rice production – the others being low rainfall and rodent damage – *C. odorata* was considered a desirable fallow species by 76% of the farmers surveyed (Roder *et al.* 1997). Such desirability as a fallow cover has also been reported in some parts of Africa (Goodall and Erasmus 1996, Norgrove *et al.* 2000).

There is currently great interest in the possible medicinal uses of non-indigenous plants (Singh and Srivastava 1999). A number of invasive species have secondary compounds that are of medicinal interest. *C. odorata* is reported to have medicinal properties (Ambika 2002), as is the neotropical, woody plant *Lantana camara*, which contains iridoid glycosides that may have potential for use as anti-cancer drugs (Ghisalberti 2000).

Lantana is on various lists of the top invasive woody weeds in the world (Cronk and Fuller 1995, Rejmánek and Richardson 1996, Binggeli *et al.* 1998). However, in addition to its medicinal properties, *L. camara* and relatives are widely valued as popular garden plants with many different brightly colored cultivars. Moreover, many agricultural benefits of *Lantana* have been reported, among them improved soil fertility and slowed soil erosion (Ghisalberti 2000). Long-term additions of *L. camara* to rice and wheat fields increased yields by up to 22% and 29%, respectively (Sharma and Verma 2000).

As with other invasive species, the negative impacts of *L. camara* on native flora appear to be variable. Gentle and Duggin (1998) documented the suppression of *Choricarpia lepropetala*, a colonizing native species of

Australian forests, by *L. camara*. However, a study in the Biligiri Rangan Hills Temple Wildlife sanctuary, Karnataka, India (Murali and Setty 2001) found no correlation between *L. camara* and species richness or stem density of other species. Interestingly, a clear negative correlation was found between these plant community metrics and the presence of another invasive, *Chromolaena odorata* (Murali and Setty 2001). Gentle and Duggin's (1998) study was experimental, while that of Murali and Setty (2001) was correlational. Additionally, the taxonomic complexity of the genus *Lantana* and the observed variation in response of different *L. camara* subspecies to biological control (Cronk and Fuller 1995) suggests that variation in interspecific competitive abilities and therefore impacts on native vegetation are likely. This is yet a further illustration of the complexity of the study of invasive species, and the contradictions, ambiguity and challenges around control.

Conclusions

Enormous challenges face managers who are confronted with developing programs to restore areas that have been invaded by introduced plants. However, an experimental approach will often help the choice among alternative, available procedures, and it may make management more efficient. There is help available in pamphlets and websites such as the Society for Ecological Restoration's *Primer on Ecological Restoration*. Development of databases is necessary to facilitate the transfer of information. A careful examination of the successes in one program may provide considerable useful information to those involved in other programs. Any large program for managing public lands will benefit from generating public awareness and promoting public involvement. In Chapter 7 we outlined a number of different ways of measuring success in biological control. These measures also apply to other types of control programs and involve such factors as educating politicians, discovering scientific information on biological interactions, and developing useful protocols. Finally, because techniques for predicting which species will become invasive will never be perfect, the best way to prevent problems in the future is to develop far more rigid quarantine procedures. In addition, we recommend the development of a regulatory and legal system that places more responsibility for future problems on those importing plants to new areas.

10 · Genetically modified plants and final conclusions

Genetically modified plants: another time bomb?

In Chapter 1 we described what Naylor (2000) calls a time bomb. This is the prolonged period of low density observed for many species before they begin to rapidly increase (Figures 1.2 and 2.8). The situation with genetically modified plants may represent another invasion time bomb. The production and release into uncontrolled environments of genetically modified crop plants (GMOs) is another frontier for invasion ecology. It might be that modification of plants could increase their invasiveness and therefore produce super weeds. More likely is the invasion of genetic material into new continents or into natural ecosystems. The fundamental assumption underlying the technology of genetic modification, is that genes from other organisms, introduced by bacteria to target plant species, will direct the production of (useful) proteins that are not normally synthesized by that plant. Currently little peer-reviewed research exists about the impacts of releasing these essentially non-native organisms into the environment. One reason for this is that it is difficult to obtain modified plants from their industrial creators, prior to their release onto the agricultural market. Once the GMOs have been released, research is more feasible, but it may then be too late to act if problems are revealed.

Many agriculture crops have close relatives among native plants. For example squash, *Cucurbita pepo*, has been modified through transgenes to be resistant to viral disease. This species occurs outside cultivation, and interbreeds with native subspecies (Parker and Kareiva 1996). It is not known whether viral disease has a role in the population dynamics of the naturally occurring gourds, but gene flow is likely to occur. Similarly, cultivated radish, *Raphanus sativus*, hybridizes with wild radish, *R. raphanistrum*, and based on flower color, crop genes are able to persist in wild populations for at least 3 years (Snow *et al.* 2001). Thus, if cultivated radish is genetically modified for resistance to disease, herbivores, herbicides or environmental stress, these genes will most likely move into wild radish in surrounding areas.

Several examples exist of gene transfers among plant varieties. While still controversial, transgenic DNA is reported to have moved into the traditional races of maize in Mexico (Quist and Chapela 2001, 2002). Canola, or oil seed rape, is another crop with closely related wild relatives with which it can potentially cross. Like its wild relatives, Canola has maintained weak seed dormancy. This means that seeds regularly persist in the seed bank and recruit as volunteer plants the next year (Pekrun *et al.* 1998). 'Roundup Ready' Canola is widely planted in the Prairie Provinces of Canada and herbicide-resistant volunteer plants are becoming a problem. Seeds can be spread widely when the crop is transported and can carry herbicide resistance genes to new areas and even new continents. Three types of herbicide-resistant Canola have been grown in Canada: those resistant to glyphosate, glufosinate and imidazolinone. Volunteers from the three varieties can persist in fields and gene flow appears to be occurring among them causing plants with multiple resistance. This 'gene stacking' means that farmers have reduced options for herbicide control of volunteer plants (Royal Society of Canada 2001). Canola is able to hybridize with at least nine related taxa (Stewart *et al.* 1997). The distribution of herbicide-resistant cultivars provides considerable opportunity for geneflow from Canola to wild relatives. This genetic mixing could create herbicide-resistant weeds. A recent study of movement of herbicide resistance among commercial Canola fields in Australia shows that pollen can carry genes for resistance at least 2.5 km. However, the frequency of transfer among fields was low, <0.02% (Rieger *et al.* 2002). Whether genes once transferred to other plant species will spread depends on their impacts on the fitness of the recipient plants. If they reduce the growth and reproduction of plants, they will be selected out of the population. If the new genes confer positive characteristics, plants carrying them will increase in frequency.

Ironically, this kind of 'genetic pollution' is the reason why the North American native tree, red mulberry, *Morus rubra*, is listed as endangered by the Canadian Committee on the Status of Endangered Wildlife in Canada (COSEWIC 2002). Pollen from the much more vigorously growing introduced white mulberry, *M. alba*, swamps the red mulberry pollen and many offspring are hybrids. There are, no doubt, many other unrecognized instances of this phenomenon. Kendle and Rose (2000) pointed out that in Britain, cultivated fritillary, *Fritillaria meleagris*, is frequently grown near wild populations in Oxfordshire, and the appearance of its flowers indicates that pollen mixing between wild and cultivated forms is occurring. The extent of the ecological impact of this genomic contamination should be investigated in an ecological framework.

Commoner (2002) pointed out that there is little information on the extent to which disruptions occur through the interactions of introduced genes and the genetic background of the modified plant. At this time, biotechnology companies are not required to show that the host plant growing in the field is actually producing a protein with the same amino acid sequence as that of the gene initially introduced. Commoner suggests that crop plants should be monitored over successive generations. Regal (1986), a pioneer in the field, reviewed various models for assessing the potential impacts of GMOs. He proposed that theories of biological invasions by non-native species could provide a model for evaluation of the possible effects of the release of genetically modified organisms.

Regulations of the US Department of Agriculture state that a transgenic plant should be no more invasive than its unmodified progenitor (USDA 1993 cited in Parker and Kareiva 1996). Parker and Kareiva (1996) review cases where these comparisons have been made and conclude that high variability makes it difficult to estimate the rates of increase in populations of transgenics and their progenitors. They feel that more care must be taken with testing cases in which the transgenic plants have gained some special protection against herbivores or diseases that could cause them to be competitively superior to unmodified plants.

A recently released report by the American National Academy of Sciences on environmental impacts of transgenic plants (NAS 2002b) concludes that introducing genes creates no new types of environmental risk and is equivalent to genetic modifications using conventional breeding. However, it suggests that the introduction of plants with specific traits that might modify their interactions with other organisms (e.g. herbivore or disease resistance) can pose environmental risks. They recommend risk assessments in these situations to evaluate competitive interactions between modified and unmodified plants. Revealing the ecological ramifications of genetically modified plants will take time. GM technology will certainly spread faster than ecological research on its consequences can be carried out. The many modifications that can be created, such as disease and herbivore resistance, and yield enhancement, will have complex influences on the fitness of plants in the wild. The rates of increase of modified and progenitor plants will differ with the environmental conditions. Furthermore, genetic mixing through cross-pollination will vary with the species and with the distribution of related species in the surrounding environment. The flow of inserted genes from crops to other species is yet another level of biological invasion that cannot be ignored.

It is doubtful that genetically modified crop plants will become broadly invasive, but the spread of their genes to other plants is a real possibility. This problem calls further attention to the global gene flow that is associated with the movement of plants. The ecological genetics of plant invasions will be an increasingly interesting and important area of study.

Some concluding remarks

Plant species, and perhaps more importantly plant diseases, are continually being carried to new environments. Some of these become established and some have major influences on plant communities. The latter can become 'keystone species' and may change the community permanently. This problem is not going to disappear. Like Pogo of the famous comic strip, 'We have met the enemy and it is us.' The most serious problems with non-native plants have involved *intentional* introductions.

The introduction of plants to new continents, ecosystems or communities creates an experimental arena in which the interactions among species can be studied. The ability of species to invade established communities reveals secrets about the patterns of species interactions. However, a consistent pattern that appears in many chapters of this book is that results of studies are both highly variable and dependent on conditions. For example, simply counting the number of species in a plot and relating this to its invasibility is not sufficient to reveal the varying influences of different functional groups. An interesting area for further study is the interactions between invasive species, and the growth and reproductive patterns of native plants already in the communities being invaded. Various theoretical models have been presented here on how resources and their variation should influence the invasibility of communities (Chapter 3). Further experimentation on how resources influence community invasibility should be done. These experiments may indicate ways in which communities could be made more resistant to invasion and thus provide useful results for land managers practising integrated weed management (Chapter 9).

One of the most exciting, and still little studied, aspects of the ecology of plant invasions is the interactions and evolution of plants and soil microbes. Will these be shown to be crucial modifiers of plant competition?

While interactions involving microscopic organisms are challenging to study, they can no longer be ignored by plant ecologists.

The introductions of insects and diseases as biological control agents are experiments on a grand scale. The evolutionary and ecological interactions between plants and a subset of their natural enemies in a new environment allows us to tease apart the varying characteristics of natural enemies. Why do some have an impact on host density and others do not? It is surprising that, in many cases, a single agent can reduce the population density of its host, and yet many agents cause considerable damage to a host without influencing its population density. Several hypotheses exist (Chapter 7) to explain the tolerance and susceptibility of plants to herbivores and diseases. These are waiting to be tested by new researchers. Just as plants move around the world, so should researchers. There is a great need for more comparative studies of plants on different continents.

Time after time we found reports in the literature that end with the hope that a biological control agent can be found that will reduce the vigor and impact of an introduced weed (Chapter 9). This demand suggests that the future is bright for biological control. The history of biological control contains examples of good science and good luck. The challenges are greater than ever, as are the opportunities for exciting ecological studies and rewarding successes. There is general agreement that more evaluation of the impacts of insects on host populations is required. Funding for post-release studies must be part of every program. The literature is rich with ideas and hypotheses on how biological control agents interact with their hosts. There are also new pressures to study the ecological ramifications of releasing exotic agents (Chapter 7). The potential efficacy of agents will become more important as we strive to be more parsimonious in selecting control agents for release. To be useful biological control models must incorporate the ability of plants to compensate for increased damage from insects. It is essential that modeling exercises be accompanied by good field measurements (Chapter 8).

Students today have ready access to a wide range of statistical and modeling techniques, but there is no substitute for counting, measuring, and observing what is happening in the field. The shapes of the relationships between density, and mortality and fecundity, are crucial to the output of theoretical models. It is important to keep in mind that these shapes can change with changing conditions in the field.

The literature on introduced species is increasing rapidly. Staying on top of all the advances is difficult because of their sheer quantity. However, the availability of information through electronic journals and other web-based resources has increased the ease with which this knowledge can be accessed. We hope that we have successfully summarized some of the background literature on which future work can be built. 'Invasion ecology' is an exciting interface where basic and applied ecology meet, and offers many career opportunites.

Some students of ecology will go on to have careers as land managers. With these positions come the pressures of dealing with and balancing the interests of the public and the demands of government agencies. Managers face similar challenges in different parts of the world and although the details of land management problems will vary, some of the underlying principles for responding to introduced plants will be the same. Obtaining data on the ecological and economic impacts of perceived weed problems should be a first step. Is the new plant species really the core problem, or are other disturbances the real culprits? A carefully planned program with clearly outlined objectives and criteria for evaluation are important for success. Communication and education are dimensions that should not be overlooked. Applied programs require both sound biology and effective public relations.

Much of the emphasis on introduced plants and their control is in the Neo-Europes. However, tropical countries in many parts of the world have devastating problems with both introduced aquatic and terrestrial weeds (Cronk and Fuller 1995). In these areas, reduction of food production by weed invasion is a major setback. Resources are likely to be limited for control programs. As Binggeli *et al.* (1998) explain, the same 'problem' species are found in many different countries, and relevant information is lodged in a wide variety of often inaccessible places. The challenge is to communicate results on problem species to people working at more local scales.

Clearly considerable variation exists in attitudes towards introduced plants. While it is a mistake to treat all plant introductions as bad, when one looks at how some ecosystems have been changed by invasive species, it is difficult to remain neutral. However, many plant species are not even recognized by the public as being exotic. The issue of plant introductions and control will thus be low on the public agenda. The ecological characteristics and the origins of plant species should be important factors in the design of landscape plantings. Increased awareness about the potential invasiveness of plants should be a component of education programs of botanical gardens and arboreta. These sources of exotic species should also develop policies to insure that they are not a cause of future problems.

Finally, the quest for clear predictors of plant invasiveness continues. There remains the question of just how good such procedures need to be. A success rate of 85% for current classifications seems pretty good (Chapter 4). However, the diversity of the life history characteristics of plants makes it almost impossible to accurately predict the potential for invasion of all introduced species. Many challenges remain.

Appendix – Some tools for studying plant populations

Introduction

The future of invasion ecology will depend on the use of good quantitative techniques. A common theme throughout this book is that data should be collected to evaluate both the extent of problems and the success of control procedures. We have included in this section a brief overview of three areas that are particularly important in the ecology of invasions: population sampling methods, measuring species diversity, and a general description of global positioning systems GPS and geographic information systems GIS. This is meant only as an introduction and we refer to other sources of more detailed coverage.

Sampling methods

The plant population ecologist has a real advantage over the animal or microbial ecologist in that, to a large degree, the target organisms stand still and can be seen and counted and reproduction and survival estimated. Of course, there are challenges with plants such as purple loosestrife, *Lythrum salicaria*, for which an individual may produce over a million seeds. Counting exceeds the limits of even the most patient graduate student. Sampling procedures require careful consideration and must be related to the biological question being asked. Books on ecological methods, e.g. Krebs (1999), are a godsend to the applied ecologist and new packages of statistical programs greatly simplify analysis. We will provide only a general overview here of techniques and considerations that influence sampling designs for plant ecology.

First, sampling must be done efficiently and this puts constraints on what can be done. In Table A.1 we outline the goals and procedures for accomplishing the goals that should precede the design of a sampling program for a study of plant population ecology.

Selecting vegetation types and representative stands will be influenced by such factors as the scale of the project, the size of the budget, the

Table A.1. *Steps in the development of a
sampling procedure for a plant population study*

Goal	Procedure
Select vegetation types	Map vegetation
Select representative stands	Define characteristic
Select sampling strategy	Completely random
	Stratified random
	Complete systematic
	Systematic-random
Select criteria to measure	Production
	Cover
	Density
	Frequency
Select sampling units	Point
	Line
	Quadrat

After Mesdaghi (2002).

availability of maps and aerial photos or GIS databases, and perhaps the presence of permanent plots or exclosures established for another study. This is, however, an important element that will determine if the scale of the study is sufficiently large to draw reliable conclusions. For example, will the scale be large enough to monitor the spread of a colonizing species? Or will the number of sites be sufficiently spaced to measure the impact of a biological control agent without control sites being overwhelmed by the released agent?

The different types of sampling methods are outlined in Figure A.1. While completely random designs may be advantageous for statistical analysis, for the field ecologist this may not provide the best solution. Choosing truly random sites can be difficult and time consuming and not all sites in a potential study area are accessible. Random sampling may require a larger number of plots to cover the heterogeneity adequately and may underestimate rare patches of vegetation that may be ecologically relevant.

By stratifying the samples among areas of the field known to differ in some way (Figure A.1C), the heterogeneity can be represented while still satisfying the requirement of random sampling. A complete systematic design involving uniform spacing of quadrats along transects

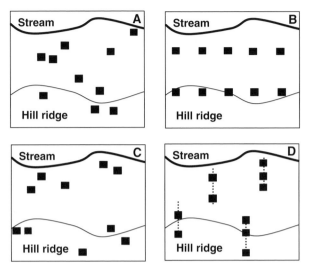

Figure A.1. Sampling designs. (A) completely random, (B) completely systematic (C) stratified random (D) systematic-random. Modified from Barbour *et al.* (1999) and Mesdaghi (2002).

chosen to represent the obvious heterogeneity of the field (Figure A.1B; see Figure 5.5 for an example based on field samples) differs from the stratified sampling procedure in not meeting the requirement of randomness required for statistical analysis, because once the starting point of the transect is chosen, subsequent sample locations will be determined. Systematic-random sampling (Figure A.1D) achieves the goal of sampling randomly and covering the heterogeneity of the field.

Point sampling for measuring cover, basal area or percent sward

Points are dimensionless quadrats. Plant cover can be estimated by dropping points vertically and recording the species of plant touched by the point as it is lowered to the ground. Cover is estimated by the proportion of pin touches of plants in general or plant species in particular.

$$\% \text{ cover} = \frac{\text{Number of pin contacts to plants or to Species A in one vertical drop}}{\text{Total number of pin drops}} \times 100$$

Plant sward refers to the proportion of the canopy thickness made up by a particular plant species, and this can be estimated by the number

Table A.2. *A ranking system for estimating percent cover of plant species in quadrat samples as recommended by Daubenmire (1959) and after Mesdaghi (2002)*

Cover class	Range of % cover	Midpoint %
1	0–1	0.5
2	1–5	3.0
3	5–25	15
4	25–50	37.5
5	50–75	62.5
6	75–100	87.5

of hits to a particular plant species by the pin as it moves through the canopy.

$$\% \text{ sward} = \frac{\text{Number of contacts with Species A}}{\text{Total number of contacts}} \times 100$$

Point samples are usually done with a single pin that is dropped from a string or with a frame through which a long pin can be dropped. This method is good for low-lying vegetation such as grasslands.

Quadrat sampling

Quadrat sampling is probably the most basic tool of the plant ecologist. It can be used to estimate cover, density, or frequency of plant species. Cover is the percentage of the quadrat covered by a particular species. One of the best ways for estimating cover is by ranking the cover class. A widely used system is the ranking scale recommended by Daubenmire (1959) (Table A.2).

Quadrats can also be used to estimate density, individuals per area; dominance, basal area or canopy cover per area; and frequency, proportion of quadrats with a particular species. These relationships are given in Box A.1.

Quadrat sampling is a basic tool in plant ecology and it is very important to decide what size and shape of quadrats are most efficient for a particular study and how many samples need to be taken.

Size and shape of sampling units

The size of sampling units will depend on the size and density of the plants being studied. Because plants have life stages that differ greatly in size from

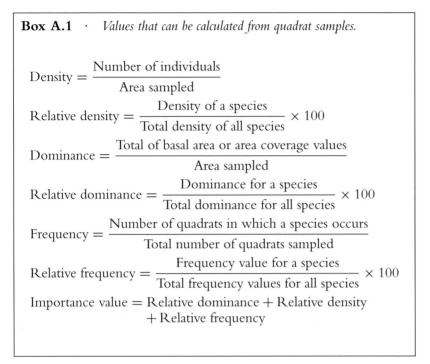

Box A.1 · *Values that can be calculated from quadrat samples.*

$$\text{Density} = \frac{\text{Number of individuals}}{\text{Area sampled}}$$

$$\text{Relative density} = \frac{\text{Density of a species}}{\text{Total density of all species}} \times 100$$

$$\text{Dominance} = \frac{\text{Total of basal area or area coverage values}}{\text{Area sampled}}$$

$$\text{Relative dominance} = \frac{\text{Dominance for a species}}{\text{Total dominance for all species}} \times 100$$

$$\text{Frequency} = \frac{\text{Number of quadrats in which a species occurs}}{\text{Total number of quadrats sampled}}$$

$$\text{Relative frequency} = \frac{\text{Frequency value for a species}}{\text{Total frequency values for all species}} \times 100$$

$$\text{Importance value} = \text{Relative dominance} + \text{Relative density} + \text{Relative frequency}$$

the seed or seedling stage to reproducing plants, the optimum sample unit may change over different stages of the life history. The optimum quadrat size will be that which (1) gives the best statistical precision, e.g. the lowest standard error or confidence interval, (2) gives the most efficient answer to the question, and (3) is the most logistically feasible (Krebs 1999). A problem that always arises in quadrat sampling is what to do with plants on the edges of the quadrat. Edge effects can cause overestimates of plant abundance (Wiegert 1962) and thus must be considered. The influence of edge effects can be measured by using quadrats of different sizes and comparing the resulting estimates of plant abundance. Small quadrat sizes that yield overestimates should be avoided. Because the distribution of plants is generally heterogeneous, long thin quadrats are often better than square quadrats for reducing the variation in abundance estimates.

In addition to the statistical efficiency of the chosen quadrat size, the relative cost of the sampling procedure will be a factor in most studies. According to Krebs (1999) the rule is to pick the quadrat size that minimizes the product of the relative cost (RC) and the relative variability (RV). Costs will involve such things as the time to pick quadrat locations,

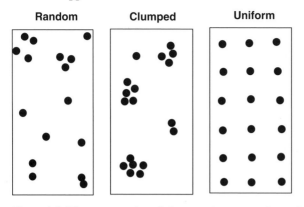

Figure A.2. Three categories of plant spacing are random, clumped and uniform.

to travel to these locations and to count the plants in the quadrats. The effectiveness, E, can be estimated as:

$$E = 1/(RT)(RV)$$

where RT = the relative time required to take a sample of a given size/minimum time to take one sample, RV = (standard deviation)^2/(minimum standard deviation)^2.

Planning ahead through a pilot study to test different quadrat sizes and shapes can be very useful in the long term.

The distribution of plants will also be relevant to sampling. Three categories of spatial patterns are random, clumped or aggregated, and uniform (Figure A.2). The underlying distribution of plants will influence both how the sampling program will be set up and how confidence limits will be calculated. The Poisson distribution fits a random pattern of plant spacing. A clumped pattern of plants is generally described by the negative binomial distribution, but aggregations of plants can be large or small with individuals within the clumps being randomly or uniformly distributed. Therefore once the pattern departs from random, a variety of distributions are possible.

The distribution of plants will influence the statistical treatment of quadrat data. Distributions can change with the developmental stage of the plant. Powell (1990b) demonstrated this by plotting individual seedlings and established diffuse knapweed plants in a 1 m × 6 m quadrat. Seedlings were dense in areas where flowering plants were sparse and vice versa (Figure A.3). Choosing a quadrat size that is efficient for adequately measuring the different life stages and along a gradient of density can be challenging.

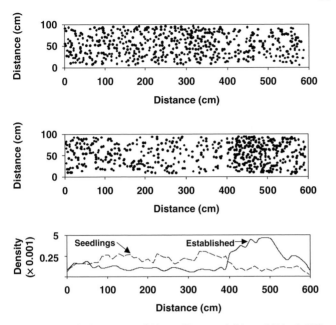

Figure A.3. The locations of (a) seedlings and (b) established diffuse knapweed plants along a 1 m × 6 m quadrat. Density estimates of the two life stages are based on non-parametric probability density estimations (Wegman 1972). After Powell (1990b).

The number of samples

The number of quadrats required will be determined by the variation among samples. Preliminary sampling can be used to estimate the required number of samples and has been described by Krebs (1999) and Cox (2001). The first step is to decide the level of precision required, e.g. ± 10% or ± 1% of the mean value. It is probably more common to want a relative level of precision in estimating plant counts, i.e. for the 95% confidence interval to be a certain percentage of the mean. The coefficient of variation is a measure of relative variability:

$$CV = s / \bar{X}$$

where s = standard deviation, $\bar{X}$ = observed mean.
From this the desired relative error can be calculated as

$$r = (t_\alpha s / \bar{X})100$$

where t_α = Student's t-value for $n - 1$ degrees of freedom for the $1 - \alpha$ level of confidence, s = standard deviation of variable.

The number of samples required (n) can be estimated by setting $t_\alpha = 2$ (a simplification for 1.96) and rearranging to obtain

$$n = (200\,\mathrm{CV}/r)^2$$

the coefficient of variation can be estimated from a subsample.

Because the distributions of plants are often clumped, quadrat counts to estimate plant population density will often fit a negative binomial distribution rather than a random distribution. In this case what is required to estimate the desired sample size is:

1 Mean value expected ($\bar{X}$),
2 Negative binomial exponent (approximate $k = \bar{X}^2/(s^2 - \bar{X})$),
3 Desired level of error (r) as a percentage,
4 Probability (α) of not achieving the desired level of error.

The variance of the negative binomial is given by

$$s^2 = \bar{X} + \bar{X}^2/k$$

and through substitution

$$n = [(100\,t_\alpha)^2/r^2][(1/\bar{X}) + (1/k)]$$

A larger number of quadrats will be required for the same precision if the pattern of a plant population fits the negative binomial rather than a Poisson.

This is just a brief overview of some of the factors that might be considered in designing a plant population study based on quadrat samples. More details should be sought from Krebs (1999) or Cox (2001). The take home message is that planning ahead can be valuable in designing and carrying out an efficient and useful population study.

Distance methods or plotless samples

An alternative to using quadrats to measure characteristics of plant populations are distance or plotless methods. In these either a random plant is chosen and the distance to its nearest neighbor measured, or random points are selected and the distance from each to the nearest plant is measured. These techniques are usually used for a single species and are illustrated in Figure A.4. Although the census zone is the area of interest, a boundary zone must be included so as to not bias results based on plants or points at the edge of the plot.

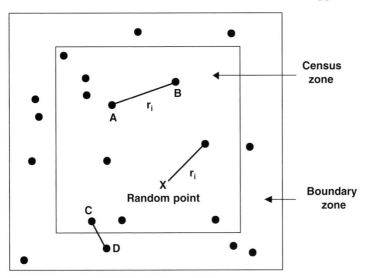

Figure A.4. Diagram of the distance measures procedure after Krebs (1999). Measurements can be from a random point (X) to the nearest plant or from a random plant (A) to its nearest neighbour (B). The nearest neighbour might be in a boundary strip outside the census zone (C to D).

If one were able to count every plant in the census plot, density could be measured directly. Often it is of interest to estimate density of plants in a large area. In this case choosing random plants becomes difficult because it is not possible to map each individual plant and choose target plants randomly. It might be possible to choose random points in the field and then select the plant closest to that point. However, Pielou (1977) has shown that system favors isolated plants. For measuring density in a large area it is necessary that sample points are well separated and therefore likely to be independent. This is where problems arise for the field ecologist (see discussion in Krebs 1999). Some type of systematic sampling is required for estimating populations over a large area and methods for this are described by Krebs (1999). These methods include the Blyth and Ripley procedure, the T-square sampling procedure, the ordered distance method, the point–quarter method and the variable-area transect method. Here we describe the variable-area transect method which is a combination of distance and quadrat methods (Parker 1979). For this, a transect of a fixed width is searched from a randomly chosen starting point to the *n*th individual of the plant species. The transect goes in only one direction and its length is determined when the *n*th

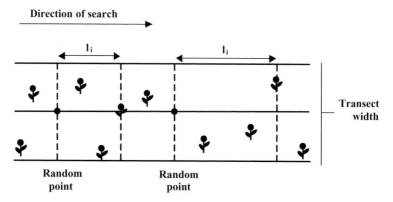

Figure A.5. Diagram of the variable-area method of estimating plant population density. This method combines quadrat and distance measurements. Random points are selected along a transect line of an arbitrary width and the length of each transect section is determined by counting plants along the transect until a fixed number of plants is reached, in this case 3 plants. A sample of 30 to 50 distances is required for a good estimate of population density.

individual is found. This procedure is diagramed in Figure A.5 and starts with deciding on the width of the transect, the number of plants to be counted between the random point and the end of the transect, and the random points along the linear transect. Starting at the first random point, x plants are counted and the length measured. The density of plants is calculated as

$$D = \frac{x(n-1)}{w\,\Sigma(l_i)}$$

where $D =$ estimate of population density, $n =$ number of random points, $x =$ number of plants from the random point to the end of transect I, $l_i =$ length of transect I searched until the xth individual was found, $w =$ width of the quadrat.

The distribution of plants in an area will bias these measurements and the relative biases of density measurements associated with these procedures is given in Krebs (1999).

The need for sampling – the need for measurement

Deciding on a sampling procedure and design can sometimes seem daunting but the complexities should not be used as an excuse for doing nothing. There are few long-term studies of plant populations of target plants in biological control programs and this reduces the information on which we can base an evaluation of its effectiveness. In addition, long-term and large-scale studies are necessary for syntheses of the invasion process. In some large programs it may be necessary for a number of different people in different areas to use the same procedures to obtain data. Simplicity and efficiency may dictate what is done to obtain the ultimate goal of comparable and long-term data. In other cases an individual researcher may have the luxury of intensive and detailed analysis.

While in general it is usually the case that bad data are worse than no data, there are situations in which population trends can give a hint as to how effective a procedure such as biological control or removing large herbivores has been. When establishing a sampling program the following should be considered: (1) what questions do I want to answer? (2) where can I get the latest advice on methods? (3) where can I get computer programs available to help with analyses? (4) given the available financial resources and time constraints can I collect sufficient data to answer these questions? (5) can a pilot study help test the feasibility of the sampling design? and (6) are there experimental procedures that would help elucidate what is happening in the population? Considering these questions at the beginning of a program can be valuable, but because in many situations involving the dynamics of invasive species, it is important to get on with it. Getting data in the early years of spread or at the beginning of a control program is important.

Measuring biodiversity in plant communities

The most obvious way to measure the biological diversity of a plant community is to go out, walk around and list the number of species present. Knowledge of the number of individuals in each species (species abundance) might also be useful, since some species will be rare and others common. While this sounds like a straightforward task, it is a complex enough proposition that many books and articles have been written about how best to measure and evaluate diversity data. Two of the most accessible books on biodiversity are by Magurran (1988) and Rosenzweig (1995). The preface of Rosenzweig (1995) is essential reading for anyone interested in science! Krebs (1999) also details the measurement of biodiversity and has programs available for calculating these. Based on these more detailed works we describe here some of the more common indices of biodiversity to give a flavor of how diversity of plant communities is measured.

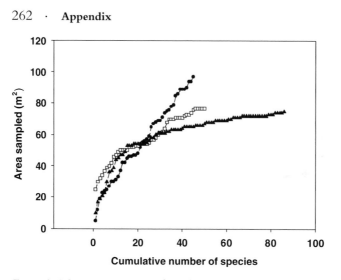

Figure A.6. Species area curves for oak-savanna plant communities in three different parks in Ontario, Canada.

Species richness

This is the simplest measurement, once you have learned to identify the species. It is the number of species in the community. Of course it will not be possible to count all of the species in a community and so we limit ourselves to particular aspects such as the diversity of flowering plants. Species richness can be the number of species per specified number of individuals or biomass. Species density is often used by botanists and is the number of species m^{-2} (Figure A.6). This is the species area curve and can also be plotted as a log–log regression or as a semi-log form.

Evenness

Not all species in a community are equally common and there will be dominant species and rare species. Some communities could have more equal abundances of most species and others could have a few dominant species and more rare species (Figure A.7). Evenness measures attempt to quantify communities with unequal representation against a hypothetical community with all equally common species.

Heterogeneity

Communities made up of equal numbers of species may still differ in how abundant the species are. Heterogeneity combines species richness

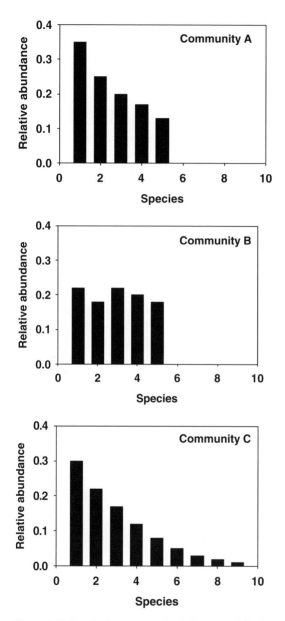

Figure A.7. Species in community A have variable abundances and therefore lower evenness as compared to the species in community B. The probability of two individuals picked from community A being the same will be lower than for community B. Therefore community B has higher heterogeneity or is more diverse. Community C has more species than communities A and B and therefore is more diverse.

and species evenness (Figure A.7). It is the measure of heterogeneity that we mostly consider to be the measure of diversity.

Measures of species richness

Rarefaction

A problem with measuring species richness is that the number of species will depend on sample size. If, for example, one is counting the number of species, it will be the case that with more individuals sampled the number of new species will increase. Rarefaction is a technique for calculating the number of species expected in each sample if the samples were of a standard size (Hurlbert 1971).

$$E(S) = \sum_{i=1}^{s} \left\{ 1 - \left[\binom{N - N_i}{n} \middle/ \binom{N}{n} \right] \right\}$$

where $E(S)$ = expected number of species, n = standardized sample size, N = total number of individuals recorded, N_i = number of individuals in the ith species. $\binom{N}{n}$ = number of combinations of n individuals that can be chosen from a set of N individuals = $N!/n!(N-n)!$

A criticism of rarefaction is that information is lost because it determines the expected number of species per sample and loses the information on the measured number of species and their abundances.

Two simple indices of diversity are Margalef's diversity index

$$D_{mg} = (S - 1)/\ln N$$

and Menhinick's index

$$D_{mn} = S/\sqrt{N}$$

where N is the total number of individuals summed over all species S.

Measures of heterogeneity

One way to present species abundance data is to do rank abundance plots. A feature of communities is that they contain comparatively few common species and large numbers of rare species. Species abundance plots have the species in rank order on the X axis and the number of individuals on the Y axis. These can be done using arithmetic scales, log–log scales, or log (y) − arithmetic (x). Krebs (1999) calls curves of species abundance using log-relative abundance (y) − arithmetic species ranks (x) 'Whittaker plots'. These plots may be all that is required to demonstrate the differences or similarities of communities.

Simpson's index

Simpson's index (D) is a measure of dominance and is weighted towards the abundances of the commonest species rather than measuring species richness. It gives the probability of any two individuals, drawn at random from an infinitely large community, belonging to different species as:

$$D = \Sigma p_i^2$$

where p_i = the proportion of individuals in the ith species. To calculate the index the form appropriate to a finite community is used:

$$D = \sum_{i=1}^{s} \left(\frac{n_i(n_i - 1)}{N(N-1)} \right)$$

where n_i = number of individuals in the ith species and N = total number of individuals. As D increases, diversity decreases and Simpson's index is usually expressed as $1 - D$ (0 is low diversity to almost 1 for high diversity) or $1/D$ (1 is low diversity to s, the number of species in the sample). It is weighted towards the most abundant species in the sample.

Shannon's index

This measure of diversity is based on information theory indices.

$$H' = -\Sigma p_i \ln p_i$$

where p_i is the proportion of individuals in the ith species. In a sample p_i is not known but is estimated as n_i/N.

The value of the Shannon diversity index usually falls between 1.5 and 3.5.

Another form of the Shannon index is:

$$N_1 = e^{H'}$$

where $e = 2.71828$ (base of natural logs), H' = Shannon function (calculated with base e logs), N_1 = number of equally common species that would produce the same diversity as H'.

This is considered to be the best heterogeneity measure for being sensitive to the abundances of the rare species in a community.

Measures of evenness

There are many indices of evenness and Krebs (1999) recommends that of Smith and Wilson (1996). It is based on the variance in abundance of the species measured over the log of the abundances. This index is:

$$E \, var = 1 - \left[\cfrac{2}{\pi \arctan \left\{ \sum_{i=1}^{s} \left(\log_e(n_i) - \sum_{j=1}^{s} \log_e(n_j)/s \right)^2 /s \right\}} \right]$$

where n_i = number of individuals in species i in sample (i = 1,2,3,4,...s), n_j = number of individuals in species j in sample (j = 1,2,3,4,...s), s = number of species in entire sample.

This index is reported to be good because it is independent of species richness and is sensitive to both rare and common species in the community.

Concluding comments of measuring diversity

Comparisons of biodiversity based on the number of species present in a habitat or community are widely used, but they suffer from a number of disadvantages.

1 They are highly dependent on sampling effort: the harder or longer you look, the more species you find.
2 They will vary between workers, depending on their ability to distinguish between species (a taxonomic expert will find more species than a non-expert).
3 The number of species present is strongly affected by habitat type and complexity. This makes comparisons between data sets from different habitats or where habitat type is uncontrolled problematic.
4 There is no statistical framework for assessing the significance of departure from expectation (are there more or fewer species present here than one might expect?).

Krebs (1999) recommends the following in regard to measurements of species diversity:

1 Construct Whittaker plots of log abundance on species rank.
2 Estimate species richness using the rarefaction method.
3 Fit the logarithmic series or the lognormal curve to the data.
4 Use the reciprocal of Simpson's index or the exponential form of the Shannon function to describe heterogeneity.

5 Use Smith and Wilson's index of evenness to estimate the evenness of the community sample.

A picture is worth a thousand words – basics of GPS and GIS

GIS, Geographic Information Systems, and GPS, Global Positioning Systems, have revolutionized the way that we think about organisms in landscapes. These systems allow data on the locations of habitats, species and communities to be relatively easily displayed in pictorial (map) format and for spatial statistics to be carried out on these data.

GPS

GPS is the acronym for Global Positioning System. This is a satellite navigation network that is owned and maintained by US Department of Defense. While the construction of the network was driven by defense needs, today there seem to be more civil applications of this technology. Fundamentally, GPS allows you to pin-point locations on earth with extraordinary precision.

A Global Positioning System is composed of three segments.

User Segment: This consists of the user and a GPS receiver, which is a specialized radio receiver. It listens to radio signals transmitted by 24 (some resources say 28) satellites orbiting the earth. The signals are then translated into position.

Space Segment: This is composed of the GPS satellites. The whole network of satellites is called a constellation. There are two different constellations: NAVSTAR and GLONASS. These constellations differ in relative positions of satellites, frequencies of signals transmitted, and the purposes they are used for.

Control Segment: This includes all the ground-based facilities that are used to monitor and control the satellites. These facilities include (1) monitoring stations. These track the satellites and determine their precise orbits. There are five such stations located at Colorado Springs, Hawaii, Ascension Island, Diego Garcia, and Kwajalein, Marshall Islands. The data collected by monitoring stations are sent to (2) the Master Control Station located at Colorado Springs. MCS receives data from the monitoring stations and checks to see if the satellites are working properly. MCS also uploads new information to the satellites via (3) the Uplink Antennas, located at Ascencion Island, Diego Garcia and Kwajalein, and use an 8-band radio link.

Depending on who the user is, GPS has two biased standards.

Standard Positioning Service (SPS): This is used by civil users world-wide without any charge or restrictions. The accuracy of signals is purposely limited through a process called Selective Availability. SPS currently provides 100 m horizontal accuracy, 156 m vertical accuracy, and 340 ns time accuracy.

Precise Positioning Services (PPS): This is the most accurate positioning, velocity, and timing information available through GPS. This service is limited to authorized US and allied Federal Governments, authorized foreign and military users, and eligible civil users. PPS signals are encrypted and are readable only by special receivers. PPS provides 22 m horizontal accuracy, 22.7 m vertical accuracy, and 200 ns time accuracy.

Applications

GPS has a variety of applications all around the world, whether it be on land, at sea or in the air. The most obvious airborne application is air traffic control. Without GPS, pilots will not be able to take off or touch down. GPS is used for navigation at sea level as well. It even tells fishermen where the best fishing spots are. Land-based applications of GPS are more versatile and diverse. They range from finding the nearest bathroom to vehicle tracking by rental companies to monitoring movement of ice sheets.

GIS

GIS is both a database as well as a collection of operations used to analyze the data. GIS is a computer system that not only records, stores, and analyzes information about features that make up this earth, but is also flexible enough to map the human body.

GIS was first built by the Canadian Government in the 1960s to analyze data collected by the Canada Land Inventory. As with the internet, GIS then spread to other governments and universities, who wished to combine various sets of data that interested them to answer various questions that concerned them. As time went by and technology improved, GIS capabilities and applications boomed. Today the applications of GIS are many and they continue to grow every day. Examples of situations where GIS can help include:

1 Researching invasive plant species and their effects on biodiversity,
2 Studying environmental changes,
3 Making legislative district boundaries,
4 Finding a location for a mega mall,
5 Finding the impact of overpopulation on health systems,
6 Designing road systems,
7 Planning emergency routes, etc., etc.

In order to use GIS, one needs specific hardware, software, data, as well as trained personnel. GIS data are the most versatile. There seems to be no limit to the kind of data this database system can handle. GIS can understand information gathered from GPS, aerial photographs, scanned maps, keyboard entry, digitized maps, conventional surveys, and so on. The system converts all geographic data into digital code and arranges it into its database.

This GIS database consists of *layers* of information. Each layer represents a particular set of geographic data. For example, one layer could include information on the downtown streets in Toronto, another layer could deal with the restaurants on those streets, another layer could be of the population demographics, and still another layer could include information on empty construction plots. Then a construction engineer can ask for these layers by going into the GIS database, combine various combinations of layers to make maps showing various patterns, and eventually help his client make a better decision as to where he/she should make a new shop in downtown Toronto. On the other hand, the Police Department can obtain some layers and decide where they need more patrolling.

Geographic Information System provides us with improved management. By linking data together and keeping them in a central place (the shared database), one department can benefit from the work of others. Data are collected once and used several times for various reasons. GIS is the spreadsheet of the new millennium. Rather than looking at columns and rows of numbers in today's spreadsheet programs, we can look at colorful 3-D maps and be better informed of the patterns that exist. Multiple scenarios can be evaluated, quickly and effectively, with GIS. In order to make better decisions, we need to be better informed, and that's what GIS does for us.

More detailed information of the use of GIS and GPS in landscape ecology can be found in Gergel and Turner (2000).

Conclusions

The study of invasive plant species, their control and the possible restoration of plant communities depends on quantitative information on the distribution, abundance and diversity of both native and introduced species. To develop general patterns of the impacts of introduced species we must know the range of potential interactions that can occur between individuals and species in the context of a heterogeneous environment. The techniques we have described in this chapter are only a subsample of those that may be relevant. Statistical techniques represent a whole additional category not considered here. The take home message is: (1) plan ahead, (2) chose the techniques that are the most efficient and yet will yield relevant data, and (3) explore the most recent developments for collecting, describing and documenting data. The combination of good data and good experiments is the best way to monitor impact and change.

References

Abrahamson, W. (1980) Demography and vegetative reproduction. In *Demography and Evolution in Plant Populations*. Solbrig, O. (ed). Oxford: Blackwell Scientific Publications, pp. 89–106.

Abrams, M., Orwig, D., and Dockry, M. (1997) Dendroecology and successional status of two contrasting old-growth oak forests in the Blue Ridge Mountains, U.S.A. *Candian Journal of Forest Research* **27**: 994–1002.

Aiken, S., Newroth, P., and Wile, D. (1979) The biology of Canadian weeds. 34. *Myriophyllum spicatum* L. *Canadian Journal of Plant Science* **59**: 201–215.

Akbay, K., Howell, F., and Wooten, J. (1991) A computer simulation model of water hyacinth and weevil interactions. *Journal of Aquatic Plant Management* **29**: 15–20.

Allen, L., Sinclair, T., and Bennett, J. (1997) Evapotranspiration of vegetation of Florida: perpetuated misconceptions versus mechanistic processes. *Soil and Crop Science Society of Florida Proceedings* **56**: 1–10.

Ambika, S. (2002) Allelopathic plants. 5. *Chromolaena odorata* (L.) King and Robinson. *Allelopathy Journal* **9**: 35–41.

Anderson, G., Delfosse, E. S., Spencer, N., Prosser, C., and Richard, R. (2000) Biological control of leafy spurge: an emerging success story. In *Proceedings of the X International Symposium on Biological Control of Weeds*. Spencer, N.R. (ed). Bozeman, MT: Montana State University, pp. 15–25.

Anderson, M. (1995) Interaction between *Lythrum salicaria* and native organisms: a critical review. *Environmental Management* **19**: 225–231.

Andres, L., and Coombs, E. (1992) Scotch broom *Cytisus scoparius* (L.) Link (Leguminosae). In *Biological Control in the U.S. Western Region: Accomplishments and Benefits of Regional Research Project W-84 (1964–1989)*. Nechols, J., Andres, L., Beardsley, J., Goeden, R., and Jackson, C. (eds). Berkeley, CA: Division of Agriculture and Natural Resources, University of California, pp. 303–305.

Anonymous (1999) Eradication of *Pennisetum setaceum* on the island of La Palma. In *Medio Ambiente Canarias, Magazine of the Regional Ministry for Territorial Policy and the Environment, Government of the Canary Islands*.

Arnold, G. W. (1995) Incorporating landscape pattern into conservation programs. In *Mosaic Landscapes and Ecological Processes*. Hansson, L., Fahrig, L., and Merriam, G. (eds). London: Chapman & Hall, pp. 309–337.

Aronson, J., Ovalle, J., Avendano, R., and Ovalle, M. (1992) Early growth rate and nitrogen fixation potential in forty-four legume species grown in an acid and a neutral soil from central Chile. *Forest Ecology and Management* **47**: 225–244.

Arroyo, M. T. K., Marticorena, C., Matthei, O., and Cavieres, O. M. (2000) Plant invasions in Chile: present patterns and future predictions. In *Invasive Species in a Changing World*. Mooney, H. A., and Hobbs, R. J. (eds). Washington, DC: Island Press, pp. 385–421.

Auld, B., Menz, K., and Monaghan, N. (1978–79) Dynamics of weed spread: implications for policies of public control. *Protection Ecology* 1: 141–148.

Austin, D. (1978) Exotic plants and their effects in southeastern Florida. *Environmental Conservation* 5: 25–34.

Bailey, K., Boyetchko, S., Derby, J., Hall, W., Sawchyn, K., Nelson, T., and Johnson, D. (2000) Evaluation of fungal and bacterial agents for biological control of Canada thistle. In *Proceedings of the X International Symposium of Biological Control of Weeds*. Spencer, N. R. (ed). Bozeman, MT: Montana State University, pp. 203–208.

Baker, B. (2001) National management plan maps strategy for controlling invasive species. *Bioscience* 51: 92.

Baker, H. (1974) The evolution of weeds. *Annual Review of Ecology and Systematics* 5: 1–24.

Barbour, M., Burk, J., and Pitts, W. (1999) *Terrestrial Plant Ecology*. Menlo Park, CA: Benjamin Cummings.

Barlow, N. (1999) Models in biological control. In *Theoretical Approaches to Biological Control*. Hawkins, B. A., and Cornell, H. (eds). Cambridge: Cambridge University Press, pp. 43–70.

Baskin, C., and Baskin, J. (1998) *Seeds*. San Diego, CA: Academic Press.

Bazely, D. R., and Jefferies, R. L. (1997) Trophic interactions in arctic ecosystems and the occurrence of a terrestrial trophic cascade. In *Ecology of Arctic Environments*. Woodin, S. J., and Marquiss, M. (eds). Oxford: Blackwell Science, pp. 183–207.

Becker, E., De La Bastide, P., Hahn, R., Shamoun, S., and Hintz, W. (2000) Molecular markers for monitoring mycoherbicides. In *Proceedings of the X International Symposium on Biological Control of Weeds*. Spencer, N. (ed). Bozeman, MT: Montana State University, 301 pp.

Beerling, D. J., Bailey, J. P., and Connolly, A. P. (1994) *Fallopia japonica* (Houtt.) Ronse Decraene. *Journal of Ecology* 82: 959–979.

Begon, M., Mortimer, M., and Thompson, D. (1996) *Population Ecology: A Unified Study of Animals and Plants*. Oxford: Blackwell Science.

Belnap, J., and Phillips, S. L. (2001) Soil biota in an ungrazed grassland: response to annual grass (*Bromus tectorum*) invasion. *Ecological Applications* 11: 1261–1275.

Benton, M. J. (1995) Diversification and extinction in the history of life. *Science* 268: 52–58.

Benton, T., and Grant, A. (1999) Elasticity analysis as an important tool in evolutionary and population ecology. *Trends in Ecology and Evolution* 14: 467–471.

Berube, D., and Myers, J. (1982) Suppression of knapweed invasion by crested wheat grass in the dry interior of British Columbia. *Journal of Range Management* 35: 459–461.

Bérubé, J., and Carisse, O. (2000) Endophytic fungal flora from eastern white pine needles and apple tree leaves as a means of biological control for white pine blister rust. In *Proceedings of the X International Symposium on Biological Control of Weeds*. Spencer, N. R. (ed). Bozeman, MT: Montana State University, 241 pp.

Billyard, E. (1996) *Relationships between landscape structure and the distribution patterns of exotic herbs.* Toronto: Department of Biology, York University, 164 pp.

Binggeli, P., Hall, J. B., and Healey, J. R. (1998) An overview of invasive woody plants in the Tropics. Bangor, Wales: School of Agricultural and Forest Sciences, 83 pp.

Blossey, B., and Nötzold, R. (1995) Evolution of increased competitive ability in invasive non-indigenous plants: a hypothesis. *Journal of Ecology* **83**: 887–889.

Blossey, B., Skinner, L., and Taylor, J. (2001) Impact and management of purple loosestrife (*Lythrum salicaria*) in North America. *Biodiversity and Conservation* **10**: 1787–1807.

Bossard, C. C., Randall, J. M., and Hoshovsky, M. C. (eds) (2000) *Invasive Plants of California's Wildlands.* The University of California Press.

Boyette, C., Walker, H., and Abbas, H. (2002) Biological control of Kudzu (*Pueraria lobata*) with an isolate of *Myrothecium verrucaria. Biocontrol Science and Technology* **12**: 75–82.

Bradshaw, A. D. (1993) Introduction: understanding the fundamentals of succession. In *Primary Succession on Land.* Miles, J., and Walton, D. H. W. (eds). Oxford: Blackwell Scientific Publications, pp. 1–3.

Braiser, C. (2001) Rapid evolution of introduced plant pathogens via interspecific hybridization. *Bioscience* **51**: 123–133.

Braithwaite, R. W., Lonsdale, W. M., and Estbergs, J. A. (1989) Alien vegetation and native biota in Tropical Australia: the impact of *Mimosa pigra. Biological Conservation* **48**: 189–210.

Brewer, L. (1995) Ecology of survival and recovery from blight in American chestnut trees (*Castanea dentata* (Marsh.) Borkh.) in Michigan. *Bulletin of the Torrey Botanical Club* **122**: 40–57.

Briese, D. (1999) Classical biological control. In *Australian Weed Management Systems.* Sindel, B. (ed). Melbourne: RG and RF Richardson Publications.

Brooks, M. L. (1999) Alien annual grasses and fire in the Mojave Desert. *Madroño* **46**: 13–19.

Brooks, M. L. (2000) Competition between alien annual grasses and native annual plants in the Mojave Desert. *American Midland Naturalist* **144**: 92–108.

Brothers, T. S., and Spingarn, A. (1992) Forest fragmentation and alien plant invasion of Central Indiana old-growth forests. *Conservation Biology* **6**: 91–100.

Broughton, S. (2000) Review and evaluation of lantana biocontrol programs. *Biological Control* **17**: 272–286.

Brown, J. H., and Lomolino, M. V. (2000) Concluding remarks: historical perspective and the future of island biogeography theory. *Global Ecology and Biogeography* **9**: 87–92.

Buchan, L., and Padilla, D. (2000) Predicting the likelihood of Eurasian watermilfoil presence in lakes, a macrophyte monitoring tool. *Ecological Applications* **10**: 1442–1455.

Buchan, L. A. J., and Padilla, D. K. (1999) Estimating the probability of long-distance overland dispersal of invading aquatic species. *Ecological Applications* **9**: 254–265.

Burdon, J., and Marshall, D. (1981) Biological control and the reproductive mode of weeds. *Journal of Applied Ecology* **18**: 49–58.

Burdon, J., Groves, R. H., and Cullen, J. (1981) The impact of biological control on the distribution and abundance of *Chondrilla juncea* in south-eastern Australia. *Journal of Applied Ecology* **18**: 957–966.

Burke, M. J. W., and Grime, J. P. (1996) An experimental study of plant community invasibility. *Ecology* **77**: 776–790.

Burton, D. (2000) Rhododendrons. In *Exotic and invasive species: should we be concerned?* Bradley, P. (ed). *Proceedings 11th Conference of the Institute of Ecology and Environmental Management*. Institute of Ecology and Environmental Management.

Caley, M., and Schluter, D. (1997) The relationship between local and regional diversity. *Ecology* **78**: 70–80.

Callaway, R. M., and Aschehoug, E. (2000) Invasive plants versus their new and old neighbors: a mechanism for exotic invasion. *Science* **290**: 521–524.

Campbell, F. (2001) The science of risk assessment for phytosanitary regulation and the impact of changing trade regulations. *BioScience* **51**: 148–153.

Canham, C. D., and Loucks, O. L. (1984) Catastrophic windthrow in the presettlement forests of Wisconsin. *Ecology* **65**: 803–809.

Carlton, J. T. (1996) Biological invasions and cryptogenic species. *Ecology* **77**: 1653–1655.

Carpenter, S., Kitchell, J. F., and Hodgson, J. (1985) Cascading trophic interactions and lake productivity. *BioScience* **35**: 634–639.

Case, T. (1990) Invasion resistance arises in strongly interacting species-rich model competition. *Proceedings of the National Academy of Sciences USA* **87**: 9610–9614.

Caswell, H. (2000a) Prospective and retrospective perturbation analyses: their roles in conservation biology. *Ecology* **81**: 619–627.

Caswell, H. (2000b) *Matrix Population Models*. Sunderland, MA: Sinauer Associates.

Caughley, G., and Lawton, J. H. (1981) Plant–herbivore systems. In *Theoretical Ecology: Principles and Applications*. May, R. M. (ed). Sunderland, MA: Sinauer Associates, pp. 132–166.

Chaboudez, P., and Sheppard, A. (1995) Are particular weeds more amenable to biological control? A reanalysis of mode of reproduction and life history. In *Proceedings of the VIII International Symposium on Biological Control of Weeds*. Delfosse, E., and Scott, R. (eds). Melbourne: DSIR/CSIRO, pp. 95–102.

Chapin, F. S. (1993) Physiological controls over plant establishment in primary succession. In *Primary Succession on Land*. Miles, J., and Walton, D. H. W. (eds). Oxford: Blackwell Scientific Publications, pp. 161–178.

Chapin, F. S., Zavaleta, E. S., Eviners, V. T., Naylor, R. L., Vitousek, P. M., Reynolds, H. L. *et al.* (2000) Consequences of changing biodiversity. *Nature* **405**: 234–242.

Chapin, F. S., Sala, O. S., Burke, I. C., Grime, J. P., Hooper, D. U., Lauenroth, W. K. *et al.* (1998) Ecosystem consequences of changing biodiversity. *BioScience* **48**: 45–52.

Charudattan, R. (1991) The mycoherbicide approach with plant pathogens. In *Microbial Control of Weeds*. TeBeest, D. (ed). New York: Chapman & Hall, pp. 24–57.

Chesson, P., and Murdoch, W. (1986) Aggregation of risk: relationships among host–parasitoid models. *American Naturalist* **127**: 696–715.

Chippendale, J. (1995) The biological control of Noogoora Burr (*Xanthium occidentale*) in Queensland: an economic perspective. In *Proceedings of the VIII International Symposium on Biological Control of Weeds*. Delfosse, E., and Scott, R. (eds). Melbourne: DSIR/CSIRO, pp. 185–192.

Clements, E. J., and Foster, M. C. (1994) *Alien Plants of the British Isles*. Botanical Society of the British Isles.

Clements, F. E. (1916) *Plant Succession: an Analysis of the Development of Vegetation*. Washington, DC: Carnegie Institute of Washington.

Clinton, W. J. (1999) Executive Order: Invasive Species. Cited in Peterson and Vieglais (2001).

Clout, M., and Lowe, S. (2000) Invasive species and environmental changes in New Zealand. In *Invasive Species in a Changing World*. Mooney, H. A., and Hobbs, R. (eds). Washington, DC: Island Press, pp. 369–383.

Cloutier, D., and Watson, A. (1990) Application of modeling to biological weed control. In *VII International Symposium on Biological Control of Weeds*. Delfosse, E. (ed). Rome, Italy, pp. 5–12.

Coleshaw, T. (ed) 2001 *The Practical Solutions Handbook for Removal of Invasive Plant Species*: FACT project/English Nature.

Collier, M., Vankat, J., and Hughes, M. (2002) Diminished plant richness and abundance below *Lonicera maackii*, an invasive shrub. *American Midland Naturalist* **147**: 60–71.

Collins, S. L. (2000) Disturbance frequency and community stability in native tallgrass prairie. *American Naturalist* **155**: 311–325.

Collins, S., Glenn, S., and Gibson, D. (1995) Experimental analysis of intermediate disturbance and initial floristic composition: decoupling cause and effect. *Ecology* **76**: 486–492.

Colton, T.F., and Alpert, P. (1998) Lack of public awareness of biological invasions by plants. *Natural Areas Journal* **18**: 262–266.

Commoner, B. (2002) Unraveling the DNA myth. In *Harper's Magazine*, February 2002, pp. 39–47.

Connell, J. H., and Slatyer, R. O. (1977) Mechanisms of succession in natural communities and their role in community stability and organization. *American Naturalist* **111**: 1119–1144.

Cory, J., and Myers, J. (2000) Direct and indirect ecological effects of biological control. *Trends Ecology and Evolution* **15**: 137–139.

COSEWIC (2002). Red mulberry, *Morus rubra*. URL http://cosewic.ec.gc.ca/eng/sct1/searchdetail_e.cfm

Costanza, R. (2001) Visions, values, valuation, and the need for an ecological economics. *BioScience* **51**: 459–468.

Couch, R., and Nelson, E. (1985) *Myriophyllum spicatum* in North America. In *Proceedings of the First International Symposium on Watermilfoil (Myriophyllum spicatum) and related Haloragaceae species*. Anderson, L. (ed). Vancouver, British Columbia: The Aquatic Plant Management Society, Inc., pp. 36–39.

Cousens, R., and Croft, A. (2000) Weed populations and pathogens. *Weed Research* **40**: 63–82.

Cousens, R., and Mortimer, M. (1995) *Dynamics of Weed Populations*. Cambridge: Cambridge University Press.

Cox, G. (2001) *General Ecology: Laboratory Manual.* New York: McGraw-Hill.

Crawley, M. J. (1983) *Herbivory, the Dynamics of Animal–Plant Interactions.* Berkeley: University of California Press.

Crawley, M. (1987) What makes a community invasible? In *Colonization, Succession, and Stability – the 26th Symposium of the British Ecological Society held jointly with the Linnean Society of London.* Gray, M., Crawley, M., and Edwards, P. (eds). Oxford: Blackwell Scientific Publications, pp. 429–453.

Crawley, M. (1989) The success and failures of weed biocontrol using insects. *Biocontrol News and Information* **10**: 213–223.

Crawley, M. (1990a) The population dynamics of plants. *Philosophical Transactions of the Royal Society of London B* **330**: 125–140.

Crawley, M. (1990b) Plant life-history and the success of weed biological control projects. In *Proceedings of the VII International Symposium on Biological Control of Weeds.* Delfosse, E. (ed). Rome: Istituto Sperimentale par la Patologia Vegetale, pp. 17–26.

Crawley, M. (1997a) Life history and environment. In *Plant Ecology.* Crawley, M. (ed). Oxford: Blackwell Science, pp. 73–131.

Crawley, M. (1997b) The structure of plant communities. In *Plant Ecology.* Crawley, M. (ed). Oxford: Blackwell Science, pp. 475–471.

Crawley, M., Harvey, P., and Purvis, A. (1996) Comparative ecology of the native and alien floras of the British Isles. *Philosophical Transactions of the Royal Society of London B* **351**: 1251–1259.

Cronk, Q., and Fuller, J. (1995) *Plant Invaders.* London: Chapman & Hall.

Crosby, A. (1986) *Ecological Imperialism: the Biological Expansion of Europe, 900–1900.* Cambridge: Cambridge University Press.

Cullen, J. (1995) Predicting effectiveness: fact and fantasy. In *Proceedings VIII International Symposium on Biological Control of Weeds.* Delfosse, E., and Scott, R. (eds). Melbourne: DSIR/CSIRO, pp. 103–109.

Cullen, J., and Delfosse, E. S. (1985) *Echium plantagineum*: catalyst for conflict and change in Australia. In *Proceedings of the VI International Symposium on Biological Control of Weeds.* Delfosse, E. (ed). Ottawa: Agriculture Canada, pp. 249–292.

Cullen, J., and Whitten, M. (1995) Economics of classical biological control: a research perspective. In *Biological Control: Benefits and Risks.* Hokkanen, H. M. T., and Lynch, J. (eds). Cambridge: Cambridge University Press, pp. 270–276.

Cullen, J., Kable, P., and Catt, M. (1973) Epidemic spread of a rust imported for biological control. *Nature* **244**: 462–464.

Daehler, C. (2001) Two ways to be an invader, but one is more suitable for ecology. *Bulletin Ecological Society of America* **82**: 101–102.

Dafni, A., and Heller, D. (1990) Invasions of adventive plants in Israel. In *Biological Invasions in Europe and the Mediterranean Basin.* di Castri, F., Hansen, A. J., and Debussche, M. (eds). Dordrecht: Kluwer, pp. 135–160.

Daily, G. (1997) Introduction: what are ecosystem services? In *Nature's Services: Societal Dependence on Natural Ecosystems.* Daily, G. (ed). Washington, DC: Island Press, pp. 3–10.

D'Antonio, C., and Vitousek, P. M. (1992) Biological invasions by exotic grasses, the grass/fire cycle, and global change. *Annual Review of Ecology and Systematics* **23**: 63–87.

Darwin, C. (1859) *On the Origin of Species*. London: Murray.

Daubenmire, R. (1959) A canopy-coverage method of vegetational analysis. *Northwest Science* **33**: 43–64.

David, P. G. (1999) Response of exotics to restored hydroperiod at Dupuis Reserve, Florida. *Restoration Ecology* **7**: 407–410.

Davis, M., and Thompson, K. (2000) Eight ways to be a colonizer; two ways to be an invader: a proposed nomenclature scheme for invasion ecology. *Bulletin of the Ecological Society of America*, July: 226–230.

Davis, M., Grime, J. P., and Thompson, K. (2000) Fluctuating resources in plant communities: a general theory of invasibility. *Journal of Ecology* **88**: 528–534.

Dawson, D. (1994) Are habitat corridors conduits for animals and plants in a fragmented landscape? A review of the scientific evidence. Peterborough, UK: English Nature, 88 pp.

Dawson, F. H., and Holland, D. (1999) The distribution in bankside habitats of three alien invasive plants in the U.K. in relation to the development of control strategies. *Hydrobiologia* **415**: 193–201.

de Jong, T., and Klinkhamer, P. (1988) Population ecology of the biennials *Cirsium vulgare* and *Cynoglossum officinale* in a coastal sand-dune area. *Journal of Ecology* **76**: 366–382.

DeAngelis, D. (1992) *Dynamics of Nutrient Cycling and Food Webs*. New York: Chapman & Hall.

Debinski, D., and Holt, R. D. (2000) A survey and overview of habitat fragmentation experiments. *Conservation Biology* **14**: 342–355.

DeLoach, C. (1990) Prospects for biological control of saltcedar (*Tamarix* spp.) in riparian habitats of the southwestern United States. In *Proceedings VII International Symposium Biological Control of Weeds*. Delfosse, E. (ed). Rome, Italy: Instituto Sperimentale per la Patologia Vegetale, pp. 307–314.

DeLoach, C., and Lewis, P. (2000) Petition to release into the field the weevil *Coniatus tamarisci* from France for biological control of *Tamarix ramosissima* and *T. parviflora* weeds of riparian areas of the western United States and northern Mexico. Temple, TX: Grassland, Soil and Water Research Laboratory, 114 pp.

Dennill, G. B., and Moran, V. C. (1989) On insect–plant associations in agriculture and the selection of agents for weed biocontrol. *Annals of Applied Biology* **114**: 157–166.

Denoth, M., Frid, L., and Myers, J. H. (2002) Multiple agents in biological control: improving the odds? *Biological Control* **24**: 20–30.

DeRouw, A. (1991) The invasion of *Chromolaena odorata* (L.) King and Robinson (ex. *Eupatorium odoratum*), and competition with the native flora, in a rain forest zone, southwest Ivory Coast. *Journal of Biogeography* **18**: 13–23.

di Castri, F. (1989) History of biological invasions. In *Biological Invasions: A Global Perspective*. Drake, J. A., Mooney, H. A., di Castri, F., Groves, R. H., Kruger, F. J., Rejmánek, M., and Williamson, M. (eds). Chichester, UK: John Wiley & Sons, pp. 1–30.

di Castri, F., Hansen, A. J., and Debussche, M. (eds) (1990) *Biological Invasions in Europe and the Mediterranean Basin*. Dordrecht: Kluwer.

Diamond, J. M. (1989) The present, past and future of human-caused extinctions. *Philosophical Transactions of the Royal Society of London B* **325**: 469–477.

Dodd, A. (1940) *The Biological Campaign Against Prickly Pear*. Brisbane: Commonwealth Prickly Pear Board.

Downey, P., and Smith, J. (2000) Demography of the invasive shrub Scotch broom (*Cytisus scoparius*) at Barrington Tops, New South Wales: insights for management. *Austral Ecology* **25**: 477–485.

Dozier, H., Gaffney, J. F., McDonald, S. K., Johnson, R. R. L., and Shilling, D.G. (1998) Cogon grass in the United States: history, ecology, impacts and management. *Weed Technology* **12**: 737–743.

Duchesne, L. C. (1994) Fire and diversity in Canadian ecosystems. In *Biodiversity, Temperate Ecosystems, and Global Change*. Boyle, J. B., and Boyle, E. B. (eds). Berlin: Springer-Verlag, pp. 247–263.

Dunn, S. T. (1905) *Alien Flora of Britain*. London: West, Newman and Co.

Dunster, K (1989). Exotic plant species management plan. Point Pelee National Park, Leamington, Ontario, Parks Canada, Ontario Region. 131 pp.

Ehler, L. E. (1995) Evolutionary history of pest–enemy associations. In *Proceedings of the VIII International Symposium on Biological Control of Weeds*. Delfosse, E., and Scott, R. (eds). Melbourne: DSIR/CSIRO, pp. 83–91.

Ehler, L. E. (1998) Invasion biology and biological control. *Biological Control* **13**: 127–133.

Ehrenfeld, J. G., and Scott, N. (2001) Invasive species and the soil: effects on organisms and ecosystems processes. *Ecological Applications* **11**: 1259–1266.

Ehrenfeld, J. G., Kourtev, P., and Huang, W. (2001) Changes in soil functions following invasions of exotic understory plants in deciduous forests. *Ecological Applications* **11**: 1287–1300.

Eldredge, N. (1997) Extinction and the evolutionary process. In *Biodiversity: An Ecological Perspective*. Abe, T., Levin, S. A., and Higashi, M. (eds). New York: Springer-Verlag, pp. 59–73.

Elton, C. (1958) *The Ecology of Invasion by Animals and Plants*. London: Meuthuen.

Enquist, B., Brown, J. H., and West, G. (1998) Allometric scaling of plant energetics and population density. *Nature* **395**: 163–165.

EPA (2002) Beneficial Landscaping: What, Why, Where and How? URL http://yosemite.epa.gov/R10/ECOCOMM.NSF/webpage/+BLWWW

Eriksson, O., and Ehrlén, J. (1992) Seed and microsite limitation of recruitment in plant populations. *Oecologia* **91**: 360–364.

Espiau, C., Riviere, D., Burdon, J. J., Gartner, S., Daclinat, B., Hasan, S., and Chaboudez, P. (1998) Host–pathogen diversity in a wild system: *Chondrilla juncea–Puccinia chondrillina*. *Oecologia* **113**: 133–139.

Evans, H. (2002) Plant pathogens for biological control. In *Plant Pathologist's Pocketbook*. Waller, J., Lenné, J., and Waller, S. (eds). Wallingford, UK: CABI Publishing, pp. 366–378.

Evans, R. D., Rimer, R., Sperry, L., and Belnap, J. (2001) Exotic plant invasion alters nitrogen dynamics in an arid grassland. *Ecological Applications* **11**: 1301–1310.

Fagan, W. F., Lewis, M. A., Neubert, M. G., and van den Driessche, P. (2002) Invasion theory and biological control. *Ecology Letters* **5**: 148–157.

Fernald, M.L. (1950) *Gray's Manual of Botany*. Portland, OR: Dioscorides Press.

Finlayson, C. M., and Mitchell, D. S. (1999) Australian wetlands: the monitoring challenge. *Wetlands Ecology and Management* **7**: 105–112.

Flanagan, G., Hills, L., and Wilson, C. (2000) The successful biological control of spinyhead Sida, *Sida acuta* [Malvaceae], by *Calligrapha pantherina* (Col: Chrysomelidae) in Australia's Northern Territory. In *Proceedings of the X International Symposium on the Biological Control of Weeds*. Spencer, N.R. (ed). Bozeman, MT: Montana State University, pp. 35–41.

Force, D. C. (1972) r- and K-strategists in endemic host–parasitoid communities. *Bulletin of the Entomological Society of America* **18**: 135–137.

Forcella, F., and Harvey, S. (1983) Relative abundance in an alien weed flora. *Oecologia* **59**: 292–294.

Forman, R. T. T., and Godron, M. (1986). *Landscape Ecology*. New York, Wiley.

Fowler, S. (2000) Trivial and political reasons for the failure of classical biological control of weeds: a personal view. In *Proceedings of the X International Symposium on Biological Control of Weeds*. Spencer, N.R. (ed). Bozeman, MT: Montana State University, pp. 169–172.

Fowler, S., Syrett, P., and Jarvis, P. (2000) Will expected and unexpected non-target effects, and the new hazardous substances and new organisms act, cause biological control of broom to fail in New Zealand. In *Proceedings of the X International Symposium on Biological Control of Weeds*. Spencer, N. R. (ed). Bozeman, MT: Montana State University, pp. 173–186.

Fox, M. D. (1990) Mediterranean weeds: exchanges of invasive plants between the five mediterranean regions of the world. In *Biological Invasions in Europe and the Mediterranean Basin*. di Castri, F., Hansen, A. J., and Debussche, M. (eds). Dordrecht: Kluwer, pp. 179–200.

Fox, M., and Fox, B. (1986) The susceptibility of natural communities to invasion. In *Ecology of Biological Invasions*. Groves, R. H., and Burdon, J. J. (eds). Cambridge: Cambridge University Press, pp 57–66.

Freckleton, R., and Watkinson, A. (1998) How does temporal variability affect predictions of weed population numbers. *Journal of Applied Ecology* **35**: 340–344.

French, K., and Eardley, K. (1997) The impact of weed infestations on litter invertebrates in coastal vegetation. In *Frontiers in Ecology: Building the Links*. Klomp, N., and Lunt, I. (eds). Oxford: Elsevier Science.

Fukami, T., Naeem, S., and Wardle, D. (2001) On similarity among local communities in biodiversity experiments. *Oikos* **95**: 340–348.

Garbelotto, M., Svihra, P., and Rizzo, D. (2001) Sudden oak death syndrome fells 3 oak species. *California Agriculture* **55**: 9–19.

Gentle, C., and Duggin, J. (1998) Interference of *Choricarpia leptopetala* by *Lantana camara* with nutrient enrichment in mesic forests on the central coast of NSW. *Plant Ecology* **136**: 205–211.

Gergel, S., and Turner, M. (2000) *Learning Landscape Ecology*. New York: Springer-Verlag.

Gerlach, J. (2001) Predicting invaders. *Trends in Ecology and Evolution* **16**: 545.

Ghisalberti, E. (2000) *Lantana camara* L. (Verbenaceae). *Fitoterapia* **71**: 467–486.

Gilpin, M. (1990) Ecological prediction. *Science* **248**: 88–89.

Gitay, H., and Noble, I. R. (1997) What are functional types and how should we seek them? In *Plant Functional Types: their Relevance to Ecosystem Properties and Global Change*. Smith, T. M., Shugart, H. H., and Woodward, F. I. (eds). Cambridge: Cambridge Unversity Press, pp. 3–19.

Gleason, H. A. (1926) The individualistic concept of the plant association. *Bulletin of the Torrey Botanical Club* **53**: 7–26.

Godron, M., and Forman, R. T. T. (1983) Landscape modifications and changing ecological characteristics. In *Disturbance and Ecosystems: Components of Response*. Mooney, H. A., and Godron, M. (eds). Berlin: Springer-Verlag, pp. 12–28.

Goeden, R. (1983) Critique and revision of Harris' scoring system for selection of insect agents in biological control of weeds. *Protection Ecology* **5**: 287–301.

Goeden, R., and Kok, L. (1986) Comments on a proposed "new" approach for selecting agents for the biological control of weeds. *Canadian Entomologist* **118**: 51–58.

Goodall, J., and Erasmus, D. (1996) Review of the status and integrated control of the invasive alien weed, *Chromolaena odorata*, in South Africa. *Agriculture, Ecosystems and Environment* **56**: 151–164.

Gordon, D. R. (1998) Effects of invasive, non-indigenous plant species on ecosystem processes: lessons from Florida. *Ecological Applications* **8**: 975–989.

Gosling, L. M., and Baker, S. J. (1989) The eradication of muskrats and coypus from Britain. *Biological Journal of the Linnean Society* **38**: 39–51.

Gould, A., and Gorchov, D. (2000) Effects of the exotic invasive shrub *Lonicera maackii* on the survival and fecundity of three species of native annuals. *American Midland Naturalist* **144**: 36–50.

Grace, J., and Tilman, D. (eds) (1990) *Perspectives on Plant Competition*. New York: Academic Press.

Greaves, M. (1996) Microbial herbicides – factors in development. In *Crop Protection Agents from Nature*. Copping, L. (ed). Cambridge: Royal Society of Chemistry, pp. 444–467.

Greaves, M., Auld, B., and Holloway, P. (1998) Formulation of microbial herbicides. In *Formulation of Biopesticides, Beneficial Microorganisms, Nematodes and Seed Treatments*. Burges, H. (ed). London: Kluwer, pp. 203–233.

Green, E., and Galatowitsch, S. (2002) Effects of *Phalaris arundinacea* and nitrate-N addition on the establishment of wetland communities. *Journal of Applied Ecology* **39**: 134–144.

Green, W. (2000) Biosecurity threats to indigenous biodiversity in New Zealand. Auckland: Report to New Zealand Government, 61 pp.

Greer, G. (1995) Economics and the biological control of weeds. In *Proceedings of the VIII International Symposium on Biological Control of Weeds*. Delfosse, E., and Scott, R. (eds). Melbourne: DSIR/CSIRO, pp. 177–184.

Grigulis, K., Sheppard, A., Ash, J., and Groves, R. H. (2001) The comparative demography of the pasture weed *Echium plantagineum* between its native and invaded ranges. *Journal of Applied Ecology* **38**: 281–290.

Grime, J. P. (1974) Vegetation classification by reference to strategies. *Nature* **250**: 26–31.

Grime, J. P. (1979) *Plant Strategies and Vegetation Processes*. Chichester, UK: John Wiley & Sons.

Grime, J., Thompson, K., Hunt, R., Hodgson, J., Cornelissen, J., Rorison, I. *et al.* (1997) Integrated screening validates primary axes of specialisation in plants. *Oikos* **79**: 259–281.

Gross, K., and Werner, P. (1982) Colonizing abilities of "biennial" plant species in relation to ground cover: implications for their distributions in a successional sere. *Ecology* **63**: 921–931.

Groves, R. (1986a) Invasion of mediterranean ecosystems by weeds. In *Resilience in Mediterranean-type Ecosystems*. Dell, B., Hopkins, A., and Lamont, B. (eds). Dordrecht: Junk, pp. 129–145.

Groves, R. H. (1986b) Plant invasions of Australia: an overview. In *Ecology of Biological Invasions: An Australian Perspective*. Groves, R. H., and Burdon, J. (eds). Canberra: Australian Academy of Science, pp. 137–149.

Hairston, N., Smith, F., and Slobodkin, L. (1960) Community structure, population control and competition. *American Naturalist* **94**: 421–425.

Hall, D. W., Currey, W. L., and Orsenigo, J. R. (1998) Weeds from other places: the Florida beachhead is established. *Weed Technology* **12**: 720–725.

Hansard (1999) Hansard (House of Commons Daily Debates) [web page]. URL http://www.publications.parliament.uk/pa/cm199899/cmhansrd/vo991103/debtext/91103–08.htm

Hara, T., van der Toorn, J., and Mook, J. (1993) Growth dynamics and size structure of shoots of *Pharagmites australis*, a clonal plant. *Journal of Ecology* **81**: 47–60.

Harper, J. (1965) Establishment, aggression, and cohabitation in weedy species. In *The Genetics of Colonizing Species*. Baker, H., and Stebbins, G. (eds). New York: Academic Press.

Harper, J. (1977) *Population Biology of Plants*. New York: Academic Press.

Harris, L. D., and Sanderson, J. (2000) The remembered landscape. In *Landscape Ecology: A Top-Down Approach*. Sanderson, J., and Harris, L. D. (eds). Boca Raton, FL: CRC Press, pp. 91–112.

Harris, P. (1974) The selection of effective agents for the biological control of weeds. *Canadian Entomologist* **105**: 1495–1503.

Harris, P. (1981) Stress as a strategy in the biological control of weeds. In *Beltsville Symposia in Agricultural Research. 5. Biological Control in Crop Production*. Papavizas, G. (ed). Beltsville, MD: Allanheld, Osmun, Totowa, pp. 333–340.

Harris, P. (1984) *Carduus nutans* L., nodding thistle and *C. acanthoides* L., plumeless thistle (Compositae). In *Biological Control Programs Against Insects and Weeds in Canada 1969–1980*. Kelleher, J., and Hulme, M. (eds). Slough, UK: Commonwealth Agricultural Bureau, pp. 115–126.

Harrison, S. (1999) Native and alien species diversity at the local and regional scales in a grazed California grass-land. *Oecologia* **121**: 99–106.

Harrison, S., and Fahrig, L. (1995) Landscape pattern and population conservation. In *Mosaic Landscapes and Ecological Processes*. Hansson, L., Fahrig, L., and Merriam, G. (eds). London: Chapman & Hall, pp. 293–308.

Hartley, K. (1985) Suppression of reproduction of woody weeds using insects which destroy flowers or seeds. In *Proceedings of the VI International Symposium on Biological Control of Weeds*. Delfosse, E. (ed). Ottawa: Agriculture Canada, pp. 749–756.

Hartley, K., and Forno, I. (1992) *Biological Control of Weeds: a Handbook for Practitioners and Students*. Melbourne: Inkata Press.

Hassell, M. P., and May, R. M. (1974) Aggregation in predators and insect parasites and its effects on stability. *Journal of Animal Ecology* **43**: 567–594.

Hazard, W. (1988) Introducing crop, pasture and ornamental species into Australia – the risk of introducing new weeds. *Australian Plant Introduction Review* **19**: 19–26.

Heaney, L. R. (2000) Dynamic disequilibrium: a long-term, large-scale perspective on the equilibrium model of island biogeography. *Global Ecology and Biogeography* **9**: 59–74.

Hector, A., Schmid, B., Beierkuhnlein, C., Caldeira, M., Diemer, M., Dimitrakopoulos, P. *et al.* (1999) Plant diversity and productivity experiments in European grasslands. *Science* **286**: 1123–1127.

Hermy, H. (1994) Effects of former land use of plant species diversity and pattern in European deciduous woodlands. In *Biodiversity, Temperate Ecosystems, and Global Change*. Boyle, J. B., and Boyle, E. B. (eds). Berlin: Springer-Verlag, pp. 123–144.

Heywood, V.H. (1989) Patterns, extents and modes of invasion by terrestrial plants. In *Biological Invasions: A Global Perspective*. Drake, J. A., Mooney, H. A., di Castri, F., Groves, R. H., Kruger, F. J., Rejmánek, M., and Williamson, M. (eds). Chichester: John Wiley & Sons, pp. 31–60.

Higgins, S. I., Richardson, D. M., and Cowling, R. M. (1996) Modeling invasive plant spread: the role of plant–environment interactions and model structure. *Ecology* **77**: 2043–2054.

Hill, R., Gourlay, A., and Fowler, S. (2000) The biological control program against gorse in New Zealand. In *Proceedings of the X International Symposium on Biological Control of Weeds*. Spencer, N. (ed). Bozeman, MT: Montana State University, pp. 909–917.

Hitchmough, J., and Woudstra, J. (1999) The ecology of exotic herbaceous perennials grown in managed, native grassy vegetation in urban landscapes. *Landscape and Urban Planning* **45**: 107–121.

Hobbs, R. J. (1989) The nature and effect of disturbance relative to invasion. In *Biological Invasions: A Global Perspective*. Drake, J. A., Mooney, H. A., di Castri, F., Groves, R. H., Kruger, F. J., Rejmánek, M., and Williamson, M. (eds). Chichester, UK: John Wiley, pp. 389–405.

Hobbs, R. J., and Atkins, L. (1988) Effect of disturbance and nutrient addition on native and introduced annuals in plant communities in the Western Australia wheatbelt. *Australian Journal of Ecology* **13**: 171–179.

Hodkinson, D., and Thompson, K. (1997) Plant dispersal: the role of man. *Journal of Applied Ecology* **34**: 1484–1496.

Hoffmann, J. (1990) Interactions between three weevil species in the biocontrol of *Sesbania punicea* (Fabaceae): the role of simulation models in evaluation. *Agriculture, Ecosystems and Environment* **32**: 77–87.

Hoffman, J., and Moran, V. (1995) Biological control of *Sesbania punicea* with *Trichapion lativentre*: diminished seed production reduces seeding but not the density of a perennial weed. In *Proceedings of the VIII International Symposium on Biological Control of Weeds*. Delfosse, E., and Scott, R. (eds). Melbourne: DSIR/CSIRO, p. 203.

Hoffmann, J., and Moran, V. (1998) The population dynamics of an introduced tree, *Sesbania punicea*, in South Africa, in response to long-term damage caused by different combinations of three species of biological control agents. *Oecologia* **114**: 343–348.

Hokkanen, H., and Pimentel, D. (1984) New approach for selecting biological control agents. *Canadian Entomologist* **116**: 1109–1171.

Hokkanen, H., and Pimentel, D. (1989) New associations in biological control: theory and practice. *Canadian Entomologist* **121**: 829–840.

Holland, D. G. (2000) Giant Hogweed and Japanese Knotweed. In *Exotic and invasive species: should we be concerned?* Bradley, P. (ed). *Proceedings 11th Conference of the Institute of Ecology and Environmental Management*. Institute of Ecology and Environmental Management.

Holmes, P. M., Richardson, D. M., Van Wilgen, B. W., and Gelderbloom, C. (2000) Recovery of South African fynbos vegetation following alien woody plant clearing and fire: implications for restoration. *Austral Ecology* **25**: 631–639.

Huenneke, L., Hamburg, S., Koide, R., Mooney, H., and Vitousek, P. (1990) Effects of soil resources on plant invasion and community structure in Californian serpentine grassland. *Ecology* **71**: 478–491.

Hurlbert, S. (1971) The non-concept of species diversity: a critique and alternative parameters. *Ecology* **52**: 577–586.

Johnstone, I., Coffey, B., and Howard-Williams, C. (1985) The role of recreational boat traffic in interlake dispersal of macrophytes: a New Zealand case study. *Journal of Environmental Management* **20**: 263–280.

Julien, M. H. (ed) (1992) *Biological Control of Weeds: A World Catalogue of Agents and their Target Weeds*, 3rd edn. Canberra: Australian Centre for International Agricultural Research.

Julien, M.H., and Griffiths, M. (eds) (1998) *Biological Control of Weeds: A World Catalogue of Agents and their Target Weeds*, 4th edn. Wallingford, UK: CAB International.

Kan-Rice, P. (2001) Oak killer found in rhododendrons. *California Agriculture* **55**: 7–8.

Kareiva, P. (1990) Population dynamics in spatially complex environments. *Philosophical Transactions of the Royal Society of London B* **330**: 175–190.

Kaufman, S., and Smouse, P. (2001) Comparing indigenous and introduced populations of *Melaleuca quinquenervia* (Cav.) Blake: response of seedlings to water and pH levels. *Oecologia* **127**: 487–494.

Kelly, D., and McCallum, K. (1995) Evaluating the impact of *Rhinocyllus conicus* on *Carduus nutans* in New Zealand. In *Proceedings of the VIII International Symposium on Biological Control of Weeds*. Delfosse, E., and Scott, R. (eds). Melbourne: DSIR/CSIRO, pp. 205–211.

Kendall, K., and Keane, R. (2001) Whitebark pine decline: infection, mortality, and population trends. In *Whitebark Pine Communities*. Tomback, D., Arno, S., and Keane, R. (eds). Washington, DC: Island Press, pp. 222–242.

Kendle, A. D., and Rose, J. E. (2000) The aliens have landed! What are the justifications for 'native only' policies in landscape plantings? *Landscape and Urban Planning* **47**: 19–31.

Kennedy, T., Naeem, S., Howe, K., Knops, J., Tilman, D., and Reich, P. (2002) Biodiversity as a barrier to ecological invasion. *Nature* **417**: 636–638.

Kinbacher, K. (2000) The tangled story of Kudzu. In *The Vulcan Historical Review*, p. 25.

Klironomos, J. (2002) Feedback with soil biota contributes to plant rarity and invasiveness in communities. *Nature* **417**: 67–70.

Kloot, P. (1991) Invasive plants of southern Australia. In *Biogeography of Mediterranean Invasions*. Groves, R. H., and di Castri, F. (eds). Cambridge: Cambridge University Press, pp. 131–144.

Klotzli, F., and Grootjans, A. (2001) Restoration of natural and semi-natural wetland systems in Central Europe: progress and predictability of developments. *Restoration Ecology* **9**: 209–219.

Kluge, R. (2000) The future of biological control of weeds with insects: no more 'paranoia', no more 'honeymoon'. In *Proceedings of the X International Symposium on Biological Control of Weeds*. Spencer, N. R. (ed). Bozeman, MT: Montana State University, pp. 459–467.

Knops, J., Tilman, D., Haddad, N., Naeem, S., Mitchell, C., Haarstad, J., *et al.* (1999) Effects of plant species richness on invasion dynamics, disease outbreaks, insect abundances and diversity. *Ecology Letters* **2**: 286–293.

Koh, S., Watt, T. A., Bazely, D. R., Pearl, D. L., Tang, M., and Carleton, T. J. (1996) Impact of herbivory of white-tailed deer (*Odocoileus virginianus*) on plant community composition. *Aspects of Applied Biology* **44**: 445–450.

Kok, L., and Surles, W. (1975) Successful biological control of musk thistle by an introduced weevil, *Rhinocyllus conicus*. *Environmental Entomology* **4**: 1025–1027.

Kolar, C. S., and Lodge, D. M. (2001) Progress in invasion biology: predicting invaders. *Trends in Ecology and Evolution* **16**: 199–204.

Kornas, J. (1988) Speirochoric weeds in arable fields: from ecological specialization to extinction. *Flora* **180**: 83–91.

Kornas, J. (1990) Plant invasions in central Europe: historical and ecological aspects. In *Biological Invasions in Europe and the Mediterranean Basin*. di Castri, F., Hansen, A. J., and Debussche, M. (eds). Dordrecht: Kluwer, pp. 19–36.

Kot, M., Lewis, M., and Van den Driessche, P. (1996) Dispersal data and the spread of invading organisms. *Ecology* **77**: 2027–2042.

Krebs, C. (1999) *Ecological Methodology*. Menlo Park, CA: Addison Wesley Longman.

Krebs, C. J. (2001) *Ecology: The Experimental Analysis of Distribution and Abundance*. San Francisco: Benjamin Cummings.

Kriticos, D., and Randall, R. (2001) A comparison of systems to analyze potential weed distributions. In *Weed Risk Assessment*. Groves, R. H., Panetta, F., and Virtue, J. (eds). Collingwood, Victoria, Australia: CSIRO, pp. 181–213.

Lande, R. (1999) Extinction risks from anthropogenic, ecological and genetic factors. In *Genetics and the Extinction of Species*. Landweber, L. F., and Dobson, A. P. (eds). Princeton, NJ: Princeton University Press, pp. 1–22.

Landolt, E. (1993) Über Pflanzenarten, die sich in den letzten 150 Jahren in der Stadt Zürich stark ausgebreitet haben. *Phytocoenologia* **23**: 651–663.

Laroche, F., and Ferriter, A. (1992) The rate of expansion of *Melaleuca* in South Florida. *Journal of Aquatic Plant Management* **30**: 62–65.

Laroche, F. B. (1998) Managing Melaleuca (*Melaleuca quinquenervia*) in the Everglades. *Weed Technology* **12**: 726–732.

Lavorel, S., Prieur-Richard, A., and Grifulis, K. (1999) Invasibility and diversity of plant communities: from patterns to processes. *Diversity and Distributions* **5**: 41–49.

Lawton, J. H. (1999) Are there general rules in ecology? *Oikos* **84**: 177–192.

Lawton, J. H., and May, R. M. (eds) (1995) *Extinction Rates*. Oxford: Oxford University Press.

Le Floc'h, E., Le Houerou, H. N., and Mathez, J. (1990) History and patterns of plant invasions in Northern Africa. In *Biological Invasions in Europe and the Mediterranean Basin*. di Castri, F., Hansen, A. J., and Debussche, M. (eds). Dordrecht: Kluwer, pp. 105–133.

Le Maitre, D., van Wilgen, B., Gelderblom, C., Bailey, C., Chapman, R., and Nel, J. (2002) Invasive alien trees and water resources in South Africa: case studies of the costs and benefits of management. *Forest Ecology and Management* **160**: 143–159.

Lehman, C. L., and Tilman, D. (2000) Biodiversity, stability, and productivity in competitive communities. *American Naturalist* **156**: 534–552.

Leishman, M., and Westoby, M. (1998) Seed size and shape are not related to persistence in soil in Australia in the same way as in Britain. *Functional Ecology* **12**: 480–485.

Levine, J. (2000) Species diversity and biological invasions: relating local process to community pattern. *Science* **288**: 852–854.

Levine, J. M., and D'Antonio, C. (1999) Elton revisited: a review of evidence linking diversity and invasibility. *Oikos* **87**: 15–26.

Levins, R. (1970) Extinction. In *Some Mathematical Questions in Biology. Lectures on Mathematics in the Life Sciences*. Gerstenhaber, M. (ed). Providence, RI: American Mathematical Society, pp. 77–107.

Lewis, M. (2000) Spread rate for a nonlinear stochastic invasion. *Journal of Mathematical Biology* **41**: 430–545.

Lewis, M., and Kareiva, P. (1993) Allee dynamics and the spread of invading organisms. *Theoretical Population Biology* **43**: 141–158.

Lockhart, C., Austin, D., and Aumen, N. (1999a) Water level effects on growth of *Melaleuca* seedlings from Lake Okeechobee (Florida, USA) littoral zone. *Environmental Management* **23**: 507–518.

Lockhart, C. S., Austin, D. F., Jones, W. E., and Downey, L. A. (1999b) Invasion of Carrotwood (*Cupaniopsis anacardioides*) in Florida Natural Areas (USA). *Natural Areas Journal* **19**: 254–262.

Lomolino, M. V. (2000) A call for a new paradigm of island biogeography. *Global Ecology and Biogeography* **9**: 1–6.

Lonsdale, W. M. (1990) The self-thinning rule: dead or alive? *Ecology* **71**: 1373–1385.

Lonsdale, W. M. (1994). Inviting trouble: introduced pasture species in Northern Australia. *Australian Journal of Ecology* **19**: 345–354.

Lonsdale, W. M. (1996) Plant population processes and weed control. In *Proceedings of the IX International Symposium on Biological Control of Weeds*. Moran, V. C., and Hoffman, J. (eds). Cape Town, SA: University of Cape Town, pp. 33–37.

Lonsdale, W. M. (1999) Global patterns of plant invasions, and the concept of invasibility. *Ecology* **80**: 1522–1536.

Lonsdale, W. M., Farrell, G., and Wilson, C. (1995) Biological control of a tropical weed: a population model and an experiment for *Sida acuta*. *Journal of Applied Ecology* **32**: 391–399.

Loope, L. L., Sanchez, P. G., Tarr, P. W., Loope, W. L., and Anderson, R. L. (1988) Biological invasions of arid land nature reserves. *Biological Conservation* **44**: 95–118.

Loreau, M., and Hector, A. (2001) Partitioning selection and complementarity in biodiversity experiments. *Nature* **412**: 72–76.

Losos, J., and Schluter, D. (2000) Analysis of an evolutionary species–area relationship. *Nature* **408**: 847–850.

Louda, S. (1998) Population growth of *Rhinocyllus conicus* (Coleoptera: Curculionidae) on two species of native thistles in prairie. *Environmental Entomology* **27**: 834–841.

Louda, S. M. (1999) Negative ecological effects of the musk thistle biocontrol agent, *Rhinocyllus conicus* Foel. In *Nontarget Effects of Biological Control*. Follett, P. A., and Duan, J. J. (eds). The Netherlands: Kluwer.

Louda, S. (2000) *Rhinocyllus conicus* – insights to improve predictability and minimize risk of biological control of weeds. In *Proceedings of the X International Symposium on Biological Control of Weeds*. Spencer, N. (ed). Bozeman, MT: Montana State University, pp. 187–194.

Louda, S. (2001) Discovering an effect of insect floral herbivory on plant population density and distribution in a "green world". *Bulletin of the Ecological Society of America*, October: 229–231.

Louda, S., Kendall, D., Connor, J., and Simberloff, D. (1997) Ecological effects of an insect introduced for the biological control of weeds. *Science* **277**: 1088–1090.

Lovett Doust, L. (1981) Population dynamics and local specialization in a clonal perennial (*Ranunculus repens*). *Journal of Ecology* **69**: 743–755.

Lovich, J. E., and Bainbridge, J. E. (1999) Anthropogenic degradation of the southern California desert ecosystem and prospects for natural recovery and restoration. *Environmental Management* **24**: 309–326.

Lozon, J. D., and MacIsaac, H. J. (1997) Biological invasions: are they dependent on disturbance? *Environmental Review* **5**: 131–141.

MacArthur, R. H., and Wilson, E. O. (1963) An equilibrium theory of insular zoogeography. *Evolution* **17**: 373–387.

MacArthur, R. H., and Wilson, E. O. (1967) *The Theory of Island Biogeography*. Princeton, NJ: Princeton University Press.

MacDonald, I., Loope, L., Usher, M., and Hamann, O. (1989) Wildlife conservation and the invasion of nature reserves by introduced species: a global perspective. In *Biological Invasions: A Global Perspective*. Drake, J. A., Mooney, H. A., di Castri, F., Groves, R. H., Kruger, F. J., Rejmánek, M., and Williamson, M. (eds). Chichester, UK: John Wiley.

Mack, R. N. (1996) Predicting the identity and fate of plant invaders: emergent and emerging approaches. *Biological Conservation* **78**: 107–121.

Mack, R. N. (2000) Assessing the extent, status, and dynamism of plant invasions: current and emerging approaches. In *Invasive Species in a Changing World*. Mooney, H. A., and Hobbs, R. J. (eds). Washington, DC: Island Press, pp. 141–168.

Mack, R., and Lonsdale, W. (2001) Humans as global plant dispersers: getting more than we bargained for. *BioScience* **51**: 95–102.

Mack, R., Simberloff, D., Lonsdale, W., Evans, H., Clout, M., and Bazzaz, F. (2000) Biotic invasions: causes, epidemiology, global consequences, and control. *Ecological Applications* **10**: 689–710.

Magurran, A. E. (1988) *Ecological Diversity and Its Measurement*. Princeton, NJ: Princeton University Press.

Mahoro, S. (2002) Individual flowering schedule, fruit set, and flower and seed predation in *Vaccinium hirtum* Thunb. (Ericaceae). *Canadian Journal of Botany* **82**: 82–92.

Maillet, J., and Lopez-Garcia, C. (2000) What criteria are relevant for predicting the invasive capacity of a new agricultural weed? The case of invasive American species in France. *Weed Research* **40**: 11–26.

Malecki, R., Blossey, B., Hight, S., Schroeder, D., Kok, L., and Coulson, R. (1993) Biological control of purple loosestrife. *BioScience* **43**: 680–686.

Maltby, L., and Mack, C. (2002) Draft report of the National Workshop on Invasive Alien Species. Ottawa: Environment Canada, 40 pp.

Mann, J. (1970) *Cacti Naturalised in Australia and Their Control.* Brisbane: Department of Lands, Queensland.

Maron, J., and Gardner, S. (2000) Consumer pressure, seed versus safe-site limitation, and plant population dynamics. *Oecologia* **124**: 260–269.

Maron, J., and Vilà, M. (2001) When do herbivores affect plant invasion? Evidence for the natural enemies and biotic resistance hypotheses. *Oikos* **95**: 361–373.

Martin, R., and Carnahan, J. (1983) A population model for Noogoora burr (*Xanthium occidentale*). *Australian Rangeland Journal* **5**: 54–62.

Mauchamp, A. (1997) Threats from alien plant species in the Galápagos. *Conservation Biology* **11**: 260–263.

May, R.M. (1973) *Stability and Complexity in Model Ecosystems.* Princeton, NJ: Princeton University Press.

McCann, K. (2000) The diversity–stability debate. *Nature* **405**: 228–233.

McCartney, J. (1999) Gardening blossoms into year-round category [web page]. URL http://www.icsc.org/srch/sct/current/index.html

McDonald, G., and Hoff, R. (2001) Blister rust: an introduced plague. In *Whitebark Pine Communities*. Tomback, D., Arno, S., and Keane, R. (eds). Washington, DC: Island Press, pp. 193–220.

McEvoy, P., and Coombs, E. (1999) Biological control of plant invaders: regional patterns, field experiments, and structured population models. *Ecological Applications* **9**: 387–401.

McEvoy, P., Rudd, N., Cox, C., and Huso, M. (1993) Disturbance, competition and herbivory effects on ragwort, *Senecio jacobaea* populations. *Ecological Monographs* **63**: 55–75.

McFadyen, R. (2000) Successes in biological control of weeds. In *Proceedings of the X International Symposium on Biological Control of Weeds*. Spencer, N. R. (ed). Bozeman, MT: Montana State University, pp. 3–14.

McFadyen, R. E. (1998) Biological control of weeds. *Annual Review of Entomology* **43**: 369–393.

McLachlan, S. M., and Bazely, D. R. (2001) Recovery patterns of understory herbs and their use as indicators of deciduous forest regeneration. *Conservation Biology* **15**: 98–110.

McNeely, J. A. (2000) The future of alien invasive species: changing social views. In *Invasive Species in a Changing World*. Mooney, H. A., and Hobbs, R. (eds). Washington, DC: Island Press, pp. 171–189.

Meijden, E.v.d., and Waals-kooi, R. E. v. d. (1979) The population ecology of *Senecio jacobaea* in a sand dune system. *Journal of Ecology* **67**: 131–153.

Meijden, E. v. d., Klinkhamer, P., de Jong, T., and van Wijk, C. (1992) Meta-population dynamics of biennial plants: how to exploit temporary habitats. *Acta Botanica Neerlandica* **41**: 249–270.

Memmott, J., Fowler, S., Paynter, Q., Sheppard, A., and Syrett, P. (2000) The invertebrate fauna on broom, *Cytisus scoparius*, in two native and two exotic habitats. *Acta Oecologia* **21**: 213–222.

Merriam, G. (1991) Corridors and connectivity: animal populations in heterogeneous environments. In *The Role of Corridors*. Saunders, D., and Hobbs, R.J. (eds). Chipping Norton, NSW: Surrey Beatty, pp. 133–142.

Mesdaghi, M. (2002) Laboratory and Field Methods for Vegetation Measurement. Gorgan, Iran: University of Agricultural Sciences and Natural Resources.

Mielke, J. (1943) White pine blister rust in western North America. New Haven, CT: Yale University, School of Forestry.

Milchunas, D. G., and Lauenroth, W. K. (1995) Inertia in plant community structure: state changes after cessation of nutrient-enrichment stress. *Ecological Applications* **5**: 452–458.

Miles, J., and Walton, D. H. W. (1993) Primary succession revisited. In *Primary Succession on Land*. Miles, J., and Walton, D. H. W. (eds). Oxford: Blackwell Scientific Publications, pp. 295–302.

Monro, J. (1967) The exploitation and conservation of resources by populations of insects. *Journal of Animal Ecology* **36**: 531–547.

Moody, M., and Mack, R. (1988) Controlling the spread of plant invasions: the importance of nascent foci. *Journal of Applied Ecology* **25**: 1009–1021.

Moran, V., and Zimmerman, H. (1991) Biological control of jointed cactus, *Opuntia aurantiaca* (Cactaceae), in South Africa. *Agriculture, Ecosystems and Environment* **37**: 5–27.

Morin, L. (1996) Different countries, several potential bioherbicides, but always the same hurdles. In *Proceedings of the IX International Symposium on Biological Control of Weeds*. Moran, V. (ed). Cape Town, SA: University of Cape Town, p. 546.

Morris, M. (1997) Impact of the gall-forming rust fungus *Uromycladium tepperianum* on the invasive tree *Acacia saligna* in South Africa. *Biological Control* **10**: 75–82.

Morrison, R. G., and Yarranton, G. A. (1974) Vegetational heterogeneity during a primary sand dune succession. *Canadian Journal of Botany* **52**: 397–410.

Mullahey, J. J., Shilling, D. G., Mislevy, P., and Akanda, R. A. (1998) Invasion of tropical soda apple (*Solanum viarum*) into the U.S.: lessons learned. *Weed Technology* **12**: 733–736.

Müller-Schärer, H., Scheepens, P., and Greaves, M. (1999) Biological control of weeds in European crops: recent achievements and future work. *Weed Research* **40**.

Mullin, B. H., Anderson, L. W. J., DiTomaso, J. M., Eplee, R. E., and Getsinger, K. D. (2000) Invasive Plant Species. US Council for Agricultural Science and Technology, 18 pp.

Murali, K., and Setty, R. (2001) Effect of weeds *Lantana camara* and *Chromelina odorata* growth on the species diversity, regeneration and stem density of tree and shrub layer in BRT sanctuary. *Current Science* **80**: 675–678.

Murdoch, W. (1990) The relevance of pest–enemy models to biological control. In *Critical Issues in Biological Control*. Mackauer, M., Ehler, L. E., and Roland, J. (eds). Andover, UK: Intercept, pp. 1–24.

Murdoch, W., and Briggs, C. (1996) Theory for biological control. *Ecology* **77**: 2001–2013.

Murdoch, W., and Stewart-Oaten, A. (1989) Aggregation by parasitoids and predators: effects on equilibrium and stability. *American Naturalist* **134**: 7–13.

Myers, J. H. (1976) Distribution and dispersal in populations capable of resource depletion: a simulation model. *Oecologia* **23**: 255–269.

Myers, J. H. (1980) Is the insect or the plant the driving force in the cinnabar moth–tansy ragwort system? *Oecologia* **47**: 16–21.

Myers, J. H. (1985) How many insect species are necessary for successful biocontrol of weeds? In *Proceedings of the VI International Symposium on Biological Control of Weeds*. Delfosse, E. (ed): Ottawa: Agriculture Canada, pp. 77–82.

Myers, J. H. (1995) Long-term studies and predictive models in the biological control of diffuse knapweed. In *Proceedings of the VIII International Symposium on Biological Control of Weeds*. Delfosse, E., and Scott, R. (eds). Melbourne: DSIR/CSIRO, pp. 221–224.

Myers, J.H. (2000) Why reduced seed production is not necessarily translated into successful biological weed control. In *Proceedings of the X International Symposium on Biological Control of Weeds*. Spencer, N.R. (ed). Bozeman, MT: Montana State University, pp. 151–154.

Myers, J. H. (2001) Predicting the outcome of biological control. In *Evolutionary Ecology: Concepts and Case Studies*. Fox, C., Roff, D., and Fairbairn, D. (eds). Oxford: Oxford University Press, pp. 361–370.

Myers, J. H., and Bazely, D. (1991) Thorns, spines, prickles, and hairs: are they stimulated by hebivory and do they deter herbivores. In *Phytochemical Induction by Herbivores*. Tallamy, D. (ed). New York: Wiley, pp. 325–344.

Myers, J. H., and Campbell, B. (1976) Distribution and dispersal in populations capable of resource depletion: a field study. *Oecologia* **24**: 7–20.

Myers, J. H., and Hosking, G. (2002) Eradication. In *Invasive Arthropods in Agriculture: Problems and Solutions*. Hallman, G. J., and C. P. Schwalbe (eds). Enfield, NH: Science Publishers, Inc.

Myers, J. H., and Risley, C. (2000) Why reduced seed production is not necessarily translated into successful biological weed control. In *Proceedings of the X International Symposium on Biological Control of Weeds*. Spencer, N. (ed). Bozeman, MT: Montana State University, pp. 569–581.

Myers, J. H., and Ware, J. (2002) Setting priorities for the biological control of weeds: What to do and how to do it. In *Proceedings, Hawaii Biological Control Workshop. Technical Report #129*. Denslow, J. E., Hight, S. D., and Smith, C. W. (ed). Honolulu: Pacific Cooperative Studies Unit, University of Hawaii, pp. 62–74.

Myers, J. H., Monro, J., and Murray, N. (1981) Egg clumping, host plant selection and population regulation in *Cactoblastis cactorum* (Lepidoptera). *Oecologia* **51**: 7–13.

Myers, J. H., Risley, C., and Eng, R. (1990) The ability of plants to compensate for insect attack: Why biological control of weeds with insects is so difficult. In *Proceedings of the VII International Symposium on Biological Control of Weeds*. Delfosse, E. (ed). Rome: Inst. Sper. Patol. Veg., pp. 67–73.

Myers, J. H., Simberloff, D., Kuris, A., and Carey, J. (2000) Eradication revisited: dealing with exotic species. *Trends in Ecology and Evolution* **15**: 316–320.

Myers, N. (1976) An expanded approach to the problem of disappearing species. *Science* **193**: 198–202.

NAS (2002a) Predicting Invasions of Nonindigenous Plants and Plant Pests. URL http://books.nap.edu/books/0309082641/html/index.html

NAS (2002b) Environmental Effects of Transgenic Plants: The Scope and Adequacy of Regulation. URL http://www.nap.edu/openbook/0309082633/html

National Invasive Species Council (2001) Meeting the Invasive Species Challenge: National Invasive Species Management Plan. Washington, DC: US Government, 89 pp.

Naylor, R. (2000) The economics of alien species invasions. In *Invasive Species in a Changing World*. Mooney, H. A., and Hobbs, R. (eds). Washington, DC: Island Press, pp. 241–259.

Neubert, M., and Caswell, H. (2000) Demography and dispersal: calculation and sensitivity analysis of invasion speed for structured populations. *Ecology* **81**: 1613–1628.

New Zealand Government (1993) Biosecurity Act. Http://www.mfe.govt.nz/laws/biosecurity.html

Nicholson, A. (1933) The balance of animal populations. *Journal of Animal Ecology* **2**: 132–178.

Noble, I., and Weiss, P. (1989) Movement and modeling of buried seed of the invasive perennial *Chrysanthemoides monilifera* in coastal dunes and biological control. *Australian Journal of Ecology* **14**: 55–64.

Norgrove, L., Hauser, S., and Weise, S. (2000) Response of *Chromolaena odorata* to timber tree densities in an agrisilvicultural system in Cameroon: aboveground biomass, residue decomposition and nutrient release. *Agriculture, Ecosystems and Environment* **81**: 191–207.

Noss, R. F. (1996) Ecosystems as conservation targets. *Trends in Ecology and Evolution* **11**: 351.

Novak, S. J., and Mack, R. N. (2001) Tracing plant introductions and spread: genetic evidence from *Bromus tectorum* (Cheatgrass). *BioScience* **51**: 114–122.

Nuzzo, V. A. (1991) Experimental control of garlic mustard (*Alliaria petiolata* (Bieb.) Cavara & Grande) in northen Illinois using fire, herbicide, and cutting. *Natural Areas Journal* **11**: 158–167.

Olckers, T., and Hill, M. (eds) (1999) *Biological Control of Weeds in South Africa (1990–1998)*. Johannesburg, SA: Entomological Society of Southern Africa.

OTA (1993) Harmful Non-Indigenous Species in the United States. Washington, DC: Office of Technology Assessment, US Congress.

Paine, R. (1966) Food web complexity and species diversity. *American Naturalist* **100**: 65–75.

Palmer, M., and Maurer, T. (1997) Does diversity beget diversity? A case study in crops and weeds. *Journal of Vegetation Science* **8**: 235–240.

Panetta, F. (1993) A system for assessing proposed plant introductions for weed potential. *Plant Protection Quarterly* **8**: 10–14.

Parendes, L., and Jones, J. (2000) Role of light availability and dispersal in exotic plant invasion along roads and streams in the H.J. Andrews Experimental Forest, Oregon. *Conservation Biology* **14**: 64–75.

Parker, I. (2000) Invasion dynamics of *Cytisus scoparius*: a matrix model approach. *Ecological Applications* **10**: 726–743.

Parker, I., and Kareiva, P. (1996) Assessing the risks of invasion for genetically engineered plants: acceptable evidence and reasonable doubt. *Biological Conservation* **78**: 193–203.

Parker, K. (1979) Density estimation by variable area transect. *Journal of Wildlife Management* **43**: 484–492.

Pastor, J., and Naiman, R. J. (1992) Selective foraging and ecosystem processes in boreal forests. *American Naturalist* **139**: 690–705.

Paynter, Q., Fowler, S., Hinz, H., Memmott, J., Shaw, R., Sheppard, A., and Syrett, P. (1996) Are seed-feeding insects of use for the biological control of broom? In *Proceedings of the IX International Symposium Biological Control of Weeds*. Moran, V., and Hoffmann, J. (eds). Cape Town, SA: University of Cape Town, pp. 495–501.

Paynter, Q., Fowler, S., Memmott, J., and Sheppard, A. (1998) Factors affecting the establishment of *Cytisus scoparius* in southern France. *Journal of Applied Ecology* **35**: 582–595.

Pekrun, C., Hewitt, J., and Hewitt, P. (1998) Cultural control of volunteer rape. *Journal of Agricultural Science* **130**: 150–163.

Pemberton, R., and Cordo, H. (2001) Potential and risks of biological control of *Cactoblastis cactorum* (Lepidoptera: Pyralidae) in North America. *Florida Entomologist* **84**: 513–526.

Pemberton, R. W. (1995) *Cactoblastis cactorum* (Lepidoptera: Pyralidae) in the United States: an immigrant biological control agent or an introduction of the nursery industry? *American Entomologist* **41**: 230–232.

Pemberton, R. W. (2000a) Naturalization pattern of horticultural plants in Florida. In *Proceedings of the X International Symposium on Biological Control of Weeds*. Spencer, N. R. (ed). Bozeman, MT: Montana State University, p. 881.

Pemberton, R. W. (2000b) Predictable risk to native plants in weed biological control. *Oecologia* **125**: 489–494.

Perrins, J., Fitter, A., and Williamson, M. (1993) Population biology and rates of invasion of three introduced *Impatiens* species in the British Isles. *Journal of Biogeography* **20**: 33–44.

Peschken, D., and McClay, A. (1995) Picking the target: a revision of McClay's scoring system to determine the suitability of a weed for classical biological control. In *Proceedings of the VIII International Symposium on Biological Control of Weeds*. Delfosse, E., and Scott, R. (eds). Melbourne: DSIR/CSIRO, pp. 137–143.

Peterson, A. T., and Vieglas, D. A. (2001) Predicting species invasions using ecological niche modeling: new approaches from bioinformatics attack a pressing problem. *Bioscience* **51**: 363–371.

Petraitis, P. S., Latham, R. E., and Niesenbaum, R. A. (1989) The maintenance of species diversity by disturbance. *Quarterly Review of Biology* **64**: 393–418.

Pettit, N. E., Ladd, P. G., and Froend, R. H. (1998) Passive clearing of native vegetation: livestock damage to remnant jarrah (*Eucalyptus marginata*) woodlands in western Australia. *Journal of the Royal Society of Western Australia* **81**: 95–106.

Pheloung, P., Williams, P., and Halloy, S. (1999) A weed risk assessment model for use as a biosecurity tool evaluating plant introductions. *Journal of Environmental Management* **57**: 239–251.

Pielou, E. (1977) *Mathematical Ecology*. New York: Wiley.

Pimentel, D., Lach, L., Zuniga, R., and Morrison, D. (2000) Environmental and economic costs of nonindigenous species in the United States. *Bioscience* **50**: 53–66.

Pimm, S. L., Russell, G. J., Gittleman, J. L., and Brooks, T. M. (1995) The future of biodiversity. *Science* **269**: 347–350.

Pitcairn, M., Woods, D., Joley, D., Turner, C., and Balciunas, J. (1999) Population buildup and combined impact of introduced insects on yellow starthistle, *Centaurea solstitialis*, in California. In *Proceedings of the X International Symposium on Biological Control of Weeds*. Spencer, N. R. (ed). Bozeman, MT: Montana State University, pp. 747–751.

Porter, W. (2001). City Gardening: Gardening a growing business but climatic predictions dubious. URL http://www.rittenhouse.ca/hortmag/Sly/City%20Gardener/Articles%202000/October2000CleanUpTime.htm

Powell, R. (1990a) The functional forms of density-dependent birth and death rates in diffuse knapweed (*Centaurea diffusa*) explain why it has not been controlled by *Urophora affinis*, *U. quadrifasciata* and *Sphenoptera jugoslavica*. In *Proceedings of the VII International Symposium on Biological Control of Weeds*. Delfosse, E. (ed). Rome: Ist. Sper. Patol. Veg. (MAF), pp. 195–202.

Powell, R. (1990b) The role of spatial pattern in the population biology of *Centaurea diffusa*. *Journal of Ecology* **78**: 374–388.

Pyke, D. (1990) Comparative demography of co-occurring introduced and native tussock grasses: persistence and potential expansion. *Oecologia* **82**: 537–543.

Quezel, P., Barbero, M., Bonin, G., and Loisel, R. (1990) Recent plant invasions in the Circum-Mediterranean region. In *Biological Invasions in Europe and the Mediterranean Basin*. di Castri, F., Hansen, A. J., and Debussche, M. (eds). Dordrecht: Kluwer, pp. 51–60.

Quist, D., and Chapela, I. (2001) Transgenic DNA introgressed into traditional maize landraces in Oaxaca, Mexico. *Nature* **414**: 541–543.

Quist, D., and Chapela, I. (2002) Reply. *Nature* **416**: 602.

Rachich, J., and Reader, R. (1999) An experimental study of wetland invasibility by purple loosestrife (*Lythrum salicaria*). *Canadian Journal of Botany* **77**: 1499–1503.

Raup, D. M. (1979) Size of the Permo-Triassic bottleneck and its evolutionary implications. *Science* **206**: 217–218.

Rea, N., and Storrs, M. J. (1999) Weed invasions in wetlands of Australia's Top End: reasons and solutions. *Wetlands Ecology and Management* **7**: 47–62.

Reader, R. (1993) Control of seedling emergence by ground cover and seed predation in relation to seed size for some old-field species. *Journal of Ecology* **81**: 169–175.

Rees, M. (1993) Trade-offs among dispersal strategies in British plants. *Nature* **336**: 150–152.

Rees, M. (1997) Seed dormancy. In *Plant Ecology*. Crawley, M. (ed). Oxford: Blackwell Science, pp. 214–238.

Rees, M., and Paynter, Q. (1997) Biological control of Scotch broom: modeling the determinants of abundance and the potential impact of introduced insect herbivores. *Journal of Applied Ecology* **34**: 1203–1221.

Reever Morghan, K. J., and Seastedt, T. R. (1999) Effects of soil nitrogen reduction on nonnative plants in restored grasslands. *Restoration Ecology* **7**: 51–55.

Regal, P. J. (1986) Models of genetically engineered organisms and their ecological impact. In *Ecology of Biological Invasions of North America and Hawaii*. Mooney, H. A., and Drake, J. A. (eds). New York: Springer-Verlag, pp. 111–129.

Regan, H. M., Lupia, R., Drinnan, A. N., and Burgman, M. A. (2001) The currency and tempo of extinction. *American Naturalist* **157**: 1–10.

Reichard, S., and Campbell, F. (1996) Invited but unwanted. *American Nurseryman*: 39–45.

Reichard, S., and Hamilton, C. (1997) Predicting invasions of woody plants introduced into North America. *Conservation Biology* **11**: 1993–1203.

Reichard, S., and White, P. (2001) Horticulture as a pathway of invasive plant introductions in the United States. *BioScience* **51**: 103–113.

Rejmánek, M. (1996) A theory of seed plant invasiveness: the first sketch. *Biological Conservation* **78**: 171–181.

Rejmánek, M. (2000) Invasive plants: approaches and predictions. *Austral Ecology* **25**: 497–506.

Rejmánek, M., and Randall, J. (1994) Invasive alien plants in California: 1993 summary and comparison with other areas in North America. *Madroño* **41**: 161–177.

Rejmánek, M., and Reichard, S. (2001) Predicting invaders. *Trends in Ecology and Evolution* **16**: 545.

Rejmánek, M., and Richardson, D. M. (1996) What attributes make some plant species more invasive? *Ecology* **77**: 1655–1661.

Reznik, S., Belokbyl'skiy, S., and Lobanov, A. (1994) Weed and herbivorous insect population densities at the broad spatial scale: *Ambrosia artemisiifolia* L. and *Zymogramma suturalis* F. (Col. Chrysomelidae). *Journal of Applied Entomology* **118**: 1–9.

Ricciardi, A., Steiner, W. W. M., Mack, R. N., and Simberloff, D. (2000) Toward a global information system for invasive species. *BioScience* **50**: 239–252.

Richards, J. (1984) Root growth response to defoliation in two *Agropyron* bunchgrasses: field observations with an improved root periscope. *Oecologia* **64**: 21–25.

Richardson, D. (1998) Forestry trees as invasive aliens. *Conservation Biology* **12**: 18–26.

Richardson, D., and Higgins, S. (1998) Pines as invaders in the southern hemisphere. In *Ecology and Biogeography of Pinus*. Richardson, D. (ed). Cambridge: Cambridge University Press, pp. 450–473.

Rieger, M., Lamond, M., Preston, C., Powles, S., and Roush, R. (2002) Pollen-mediated movement of herbicide resistance between commercial canola fields. *Science* **296**: 2386–2388.

Roder, W., Phengchanh, S., and Keobulapha, B. (1997) Weeds in slash-and-burn rice fields in northern Laos. *Weed Research* **37**: 111–119.

Roelfs, A. (1982) Effects of barberry eradication on stem rust in the United States. *Plant Disease* **66**: 177–181.

Room, P. (1983) "Falling apart" as a lifestyle: the rhizome architecture and population growth of *Salvinia molesta*. *Journal of Ecology* **71**: 349–365.

Room, P. (1990) Ecology of a simple plant–herbivore system: biological control of *Salvinia*. *Trends in Ecology and Evolution* **5**: 74–79.

Rosenzweig, M. L. (1995) *Species Diversity in Space and Time*. Cambridge: Cambridge University Press.

Rotherham, I. (2000) Himalayan balsam – the human touch. In *Exotic and invasive species: Should we be concerned? 11th Conference the Institute of Ecology and Environmental Management*. Mieem, P. (ed). Institute of Ecology and Environmental Management, pp. 41–47.

Royal Society of Canada (2001) Report of the Expert Panel on the Future of Food Biotechnology. Ottawa: The Royal Society of Canada.

Sala, O.E. (2001) Price put on biodiversity. *Nature* **412**: 34–36.

Salisbury, E. (1942) *The Reproductive Capacity of Plants*. London: G. Bell.

Sanderson, J., and Harris, L. D. (2000) Brief history of landscape ecology. In *Landscape Ecology: A Top-Down Approach*. Sanderson, J., and Harris, L. D. (eds). Boca Raton, FL: CRC Press, pp. 3–18.

Sands, D., and Schotz, M. (1985) Control or no control: a comparison of the feeding strategies of two *Salvinia* weevils. In *Proceedings of the VI International Symposium on Biological Control of Weeds*. Delfosse, E. (ed). Ottawa: Agriculture Canada, pp. 551–556.

Santha, C., Grant, W., Neill, W., and Strawn, R. (1991) Biological control of aquatic vegetation using grass carp: simulation of alternative strategies. *Ecological Modeling* **59**: 229–245.

Savidge, J.A. (1987) Extinction of an island forest avifauna by an introduced snake. *Ecology* **68**: 660–668.

Schaffner, U. (2000) Seedling establishment of invasive and non-invasive populations of sulphur cinquefoil, *Potentilla recta* L. In *Proceedings of the X International Symposium on Biological Control of Weeds*. Spencer, N. R. (ed). Bozeman, MT: Montana State University, pp. 600–601.

Schemske, D. (1984) Population structure and local selection in *Impatiens pallida* (Balsaminaceae), a selfing annual. *Evolution* **38**: 817–832.

Scherer-Lorenzen, M., Elend, A., Nollert, S., and Schuze, E.-D. (2000) Plant invasions in Germany: general aspects and impact of nitrogen deposition. In *Invasive Species in a Changing World*. Mooney, H. A., and Hobbs, R. (eds). Washington, DC: Island Press, pp. 351–368.

Schultz, S., Dunham, A., Root, K., Soucy, S., and Ginzburgh, L. (1999) *Conservation Biology with RAMAS EcoLab*. Sunderland, MA: Sinauer Associates.

Scott, N. A., Saggar, S., and McIntrosh, P. D. (2001) Biogeochemical impact of *Hieracium* invasion in New Zealand's grazed tussock grasslands: sustainability implications. *Ecological Applications* **11**: 1311–1322.

SER, Society for Ecological Restoration Science and Policy Working Group (2002) *The SER Primer on Ecological Restoration*. URL www.ser.org

Sharma, R., and Verma, T. (2000) Effect of long-term addition of lantana biomass on crop yields and N uptake in rice-wheat cropping in Himalayan acid Alfisols. *Tropical Agriculture* **77**: 71–75.

Sharov, A., and Liebhold, A. (1998) Bioeconomics of managing the spread of exotic pest species with barrier zones. *Ecological Applications* **8**: 833–845.

Shea, K., and Kelly, D. (1998) Estimating biocontrol agent impact with matrix models: *Carduus nutans* in New Zealand. *Ecological Applications* **8**: 824–832.

Shea, K., and Chesson, P. (2002) Community ecology theory as a framework for biological invasions. *Trends in Ecology and Evolution* **17**: 170–176.

Sheail, J. (1987) *Seventy-five years in Ecology: The British Ecological Society*. Oxford: Blackwell Scientific.

Sheppard, A., Cullen, J., and Aeschlimann, J. (1994) Predispersal seed predation on *Carduus nutans* (Asteraceae) in southern Europe. *Acta Oecologica* **15**: 529–541.

Silori, C., and Mishra, B. (2001) Assessment of livestock grazing pressure in and around the elephant corridors in Mudumalai Wildlife Sanctuary, south India. *Biodiversity and Conservation* **10**: 2181–2195.

Silvertown, J., and Lovett Doust, J. (1993) *Introduction to Plant Population Biology*. Oxford: Blackwell Scientific Publications.

Silvertown, J. (1983) Why are biennials sometimes not so few? *American Naturalist* **121**: 148–453.

Simberloff, D., Farr, J., Cox, J., and Mehlman, D. (1992) Movement corridors: conservation bargains or poor investments? *Conservation Biology* **6**: 493–504.

Simmons, F., and Bennett, F. (1966) Biological control of *Opuntia* sp by *Cactoblastis cactorum* in the Leeward Islands (West Indies). *Entomphaga* **11**: 183–189.

Singh, S. C., and Srivastava, G. N. (1999) Exotic medicinal plants of Lucknow district, U.P., India. *Journal of Economic and Taxonomic Botany* **23**: 223–235.

Skellam, J. (1951) Random dispersal in theoretical populations. *Biometrika* **38**: 196–218.

Smith, B., and Wilson, J. (1996) A consumer's guide to evenness indices. *Oikos* **76**: 70–82.

Smith, F. D. M., May, R. M., Pellew, R., Johnson, T. H., and Walter, K. S. (1993) Estimating extinction rates. *Nature* **364**: 494–496.

Smith, L., Ravlin, F., Kok, L., and Mays, W. (1984) Seasonal model of the interaction between *Rhinocyllus conicus* (Coleoptera: Curculionidae) and its weed host, *Carduus thoermeri* (Campanulatae: Asteraceae). *Environmental Entomology* **13**: 1417–1426.

Smith, M., and Holt, J. (1993) A population model for the parasitic weed *Striga hermonthica* (Scrophulariaceae) to investigate the potential of *Smicronyx umbrinus* (Coleoptera: Curculionidae) for biological control in Mali. *Crop Protection* **12**: 470–476.

Smith, M. D., and Knapp, A. K. (1999) Exotic plant species in a C4-dominated grassland: invasibility, disturbance, and community structure. *Oecologia* **120**: 605–612.

Snow, A., Uthus, K., and Culley, T. (2001) Fitness of hybrids between weedy and cultivated radish: implications for weed evolution. *Ecological Applications* **11**: 934–943.

Srivastava, D. (1999) Using local-regional richness plots to test for species saturation: pitfalls and potential. *Journal of Animal Ecology* **68**: 1–16.

Stadler, J., Trefflich, A., Klotz, S., and Brandl, R. (2000) Exotic plant species invade diversity hotspots: the alien flora of northwestern Kenya. *Ecography* **23**: 169–176.

Steadman, D. W. (1995) Prehistoric extinctions of Pacific Island birds: biodiversity meets zooarchaeology. *Science* **267**: 1123–1131.

Stearns, S. (1992) *The Evolution of Life-Histories*. Oxford: Oxford University Press.

Stephenson, S. (1986) Changes in a former chestnut-dominated forest after a half century of succession. *American Midland Naturalist* **116**: 173–179.

Stephenson, S., and Fortney, R. (1998) Changes in forest overstory composition on the southwest-facing slope of Beanfield Mountain in south western Virginia. *Castanea* **63**: 482–488.

Stewart, C., All, J., Raymer, P., and Ramachndran, S. (1997) Increased fitness of transgenic insecticidal rapeseed under selection pressure. *Molecular Ecology* **6**: 773–779.

Stiling, P., and Moon, D. (2001) Protecting rare Florida cacti from attack by the exotic cactus moth, *Cactoblastis cactorum* (Lepidoptera: Pyralidae). *Florida Entomologist* **84**: 506–512.

Stöcklin, J., and Fischer, M. (1999) Plants with longer-lived seeds have lower local extinction rates in grassland remnants 1950–1985. *Oecologia* **120**: 539–543.

Stohlgren, T., Schell, L., and Vanden Heuvel, B. (1999) How grazing and soil quality affect native and exotic plant diversity in Rocky Mountain grasslands. *Ecological Applications* **9**: 45–64.

Straw, N., and Sheppard, A. (1995) The role of plant dispersion pattern in the success and failure of biological control. In *Proceedings of the VIII International Symposium on Biological Control of Weeds*. Delfosse, E., and Scott, R. (eds). Melbourne: DSIR/CSIRO, pp. 161–168.

Stromme, I., Cole, D., Tansey, J., McClay, A., Richardson, C., and de Valois, J. (2000) Long-term monitoring of the impact of *Aphthona nigriscutis* on leafy spurge: the Beverly Bridge sites. In *Proceedings of the X International Symposium on Biological Control of Weeds*. Spencer, N.R. (ed). Bozeman, MT: Montana State University, p. 778.

Tatter, T., Berman, P., Gonzalez, E., Mount, M., and Dolloff, A. (1996) Biocontrol of the chestnut blight fungus, *Cryphonectria parasitica*. *Arboricultural Journal* **20**: 449–469.

Taylor, H. (2001) Reading, TV, spending time with family, gardening and fishing top list of favorite leisure-time activities [web page]. URL http://www.harrisinteractive.com/harris_poll/index.asp?PID=249

Taylor, R. L., and MacBryde, B. (1977) *Vascular Plants of British Columbia: A Descriptive Inventory*. University of British Columbia Press.

Templeton, G., and Trujillo, E. (1981) The use of plant pathogens in the biological control of weeds. In *Handbook of Pest Management in Agriculture*. Pimental, D. (ed). Boca Raton, FL: CRC Press, pp. 345–350.

Tewksbury, L., Casagrande, R., Blossey, B., Häfliger, P., and Schwarzländer, M. (2002) Potential for biological control of *Phragmites australis* in North America. *Biological Control* **23**: 191–212.

Thébaud, C., and Simberloff, D. (2001) Are plants really larger in their introduced ranges? *American Naturalist* **157**: 231–236.

Thomas, M.B., and Willis, A. (1998) Biocontrol – risky but necessary. *Trends in Ecology and Evolution* **13**: 325–329.

Thompson, K. (1994) Predicting the fate of temperate species in response to human disturbance and global change. In *Biodiversity, Temperate Ecosystems, and Global Change*. Boyle, J. B., and Boyle, E. B. (eds). Berlin: Springer-Verlag, pp. 61–76.

Thompson, K., Bakker, J., Bekker, R., and Hodgson, J. (1998) Ecological correlates of seed persistence in soil in the north-west European flora. *Journal of Ecology* **86**: 163–169.

Thompson, K., Band, S., and Hodgson, J. (1993) Seed size and shape predict persistence in soil. *Functional Ecology* **7**: 236–241.

Thompson, W. (1930) The utility of mathematical methods in relation to work on biological control. *Annals of Applied Biology* **17**: 641–648.

Tilman, D. (1997) Community invasibility, recruitment limitation, and grassland biodiversity. *Ecology* **78**: 81–92.

Tilman, D., Knops, J., Wedin, D., Reich, P., Ritchie, M., and Siemann, E. (1997) The influence of functional diversity and composition on ecosystem processes. *Science* **277**: 1300–1302.

Tilman, D., Wedin, D., and Knops, J. A. N. L. (1996) Productivity and sustainability influenced by biodiversity in grassland ecosystems. *Nature* **379**: 718–720.

Tomback, D., and Kendall, K. (2001) Biodiversity losses: the downward spiral. In *White Bank Pine Communities*. Washington, DC: Island Press, pp. 242–262.

Torres, J.-L., Sosa, V., Equihua, M., and Torres, L. (2001) On the conceptual basis of the self-thinning rule. *Oikos* **95**: 544–548.

Trabaud, L. (1990) Fire as an agent of plant invasion? A case study in the French Mediterranean region. In *Biological Invasions in Europe and the Mediterranean Basin*. di Castri, F., Hansen, A. J., and Debussche, M. (eds). Dordrecht: Kluwer, pp. 417–437.

Trombulak, S., and Frissell, C. (2000) Review of ecological effects of roads on terrestrial and aquatic communities. *Conservation Biology* **14**: 18–30.

Tuljapurkar, S., and Caswell, H. (eds) (1997) *Structured Population Models in Marine, Terrestrial and Freshwater Systems*. New York: Chapman & Hall.

Turkington, R., and Mehrhoff, L. A. (1990) The role of competition in structuring pasture communities. In *Perspectives on Plant Competition*. Grace, J., and Tilman, D. (eds). New York: Academic Press, pp. 308–335.

Turnbull, L., Crawley, M., and Rees, M. (2000) Are plant populations seed-limited? A review of seed sowing experiments. *Oikos* **88**: 225–238.

Turner, C. E., Center, T. D., Burrows, D. W., and Buckingham, G. R. (1998) Ecology and management of *Melaleuca quinquenervia*, an invader of wetlands in Florida, U.S.A. *Wetlands Ecology and Management* **5**: 165–178.

Turner, I. M., Chua, K. S., Ong, J. S. Y., Soong, B. C., and Tan, H. T. W. (1996) A century of plant species loss from an isolated fragment of lowland tropical rainforest. *Conservation Biology* **10**: 1229–1244.

Turner, I. M., and Corlett, R. T. (1996) The conservation value of small, isolated fragments of lowland tropical rain forest. *Trends in Ecology and Evolution* **11**: 330–333.

Turner, I. M., Tan, H. T. W., Wee, Y. C., Ibrahim, A. B., Chew, P. T., and Corlett, R. T. (1994) A study of plant species extinction in Singapore: lessons for the conservation of tropical biodiversity. *Conservation Biology* **8**: 705–712.

UNEP (1992) The Convention on Biodiversity. United Nations Environment Program. http://www.biodiv.org/convention/articles.asp

UNF (1999) URL (http://www.unfoundation.org/grants/4_3_invasives.asp

Ussery, J. G., and Krannitz, P. G. (1998) Control of Scot's broom (*Cytisus scoparius* (L.) Link.): the relative conservation merits of pulling versus cutting. *Northwest Science* **72**: 268–273.

van Kleunen, M. (2001) Evolution of clonal life-history of *Ranunculus reptans*. In *Mathematisch-Naturwissenschaftlichen Fakultät*. Zurich: Universität Zürich, 173 pp.

van Kleunen, M., and Fischer, M. (2001) Adaptive evolution of plastic foraging responses in a clonal plant. *Ecology* **82**: 3309–3319.

van Kleunen, M., Fischer, M., and Schmid, B. (2001) Effects of intraspecific competition on size variation and reproductive allocation in a clonal plant. *Oikos* **94**: 515–524.

Vandiver, V., Hall, D. W., and Westbrooks, R. (1992) Discovery of *Oryza rufipogon* (Poaceae: Oryzeae), new to the United States, with its implications. *Sida Contributions to Botany* **15**: 105–109.

Varley, G., and Gradwell, G. (1960) Key factors in populations studies. *Journal of Animal Ecology* **29**: 399–401.

Vilà, M., Garcia-Berthou, G., Sol, D., and Pino, J. (2001) Survery of the naturalised plants and vertebrates in peninsular Spain. *Ecologia Mediterranea* **27**: 55–67.

Vitousek, P., D'Antonio, C., Loope, L., and Westbrooks, R. (1996) Biological invasions as global environmental change. *American Scientist* **84**: 468–478.

Vitousek, P. M. (1986) Biological invasions and ecosystem properties. In *Ecology of Biological Invasions of North America and Hawaii*. Mooney, H. A., and Drake, J. A. (eds). New York: Springer-Verlag, pp. 163–176.

Vitousek, P. M. (1990) Biological invasions and ecosystem process – towards an integration of population biology and ecosystem studies. *Oikos* **57**: 7–13.

Vivrette, N. J., and Muller, C. H. (1977) Mechanism of invasion and dominance of coastal grassland by *Mesembryanthemum crystallinum*. *Ecological Monographs* **47**: 301–318.

Wagner, W. L., Herbst, D. R., and Sohmer, S. H. (1990) *Manual of the Flowering Plants of Hawai'i*, Vol. 1. Honolulu: University of Hawai'i Press and Bishop Museum Press.

Walker, L. R. (1993) Nitrogen fixers and species replacements in primary succession. In *Primary Succession on Land*. Miles, J., and Walton, D. H. W. (eds). Oxford: Blackwell Scientific Publications, pp. 249–272.

Waloff, N., and Richards, O. (1977) The effect of insect fauna on growth mortality and natality of broom, *Sarothamnus scoparius*. *Journal of Applied Ecology* **14**: 787–798.

Walters, C. (1986) *Adaptive Management of Renewable Resources*. New York: Macmillan.

Wapshere, A. (1985) Effectiveness of biological control agents for weeds: present quandaries. *Agriculture, Ecosystems and Environment* **18**: 261–280.

Wardle, D. (2001) Experimental demonstration that plant diversity reduces invasibility – evidence of a biological mechanism or a consequence of sampling effect? *Oikos* **95**: 161–170.

Wardle, D., Nicholson, K., Ahmed, M., and Rahman, A. (1995) Influence of pasture forage species on seedling emergence, growth and development of *Carduus nutans*. *Journal of Applied Ecology* **32**: 225–233.

Watkinson, A. (1985) On the abundance of plants along an environmental gradient. *Journal of Ecology* **73**: 569–578.

Watt, A. S. (1981a) A comparison of grazed and ungrazed grassland in East Anglian Breckland. *Journal of Ecology* **69**: 449–508.

Watt, A.S. (1981b) Further observations on the effects of excluding rabbits from Grassland A in East Anglian Breckland: the pattern of change and factors affecting it (1936–73). *Journal of Ecology* **69**: 509–536.

Watt, T. A. (1997) *Introductory Statistics for Biology Students*. London: Chapman & Hall.

Weber, E. (1999) Gebietsefremde Arten der Schwiezer Flora – Ausmass und Bedeutung. *Bauhinia* **13**: 1–10.

Wegman, E. (1972) Non-parametric probability density estimation. I. A summary of available methods. *Technometrics* **14**: 533–546.

Weller, D. (1987) A reevaluation of the −3/2 power rule of plant self-thinning. *Ecological Monographs* **57**: 23–43.

Welling, C., and Becker, R. (1990) Seed bank dynamics of *Lythrum salicaria* L.: implications for control of this species in North America. *Aquatic Botany* **38**: 303–309.

Westbrooks, R. G. (1991) Plant protection issues. I. A commentary on new weeds in the United States. *Weed Technology* **5**: 232–237.

Westbrooks, R. G., Otteni, L., and Eplee, R. E. (1997) New strategies for weed prevention. In *Exotic Pests of Eastern Forests*. Britton, K.O. (ed). Nashville, TN: USDA Forest Service & TN Exotic Pest Plant Council.

White, J. (1981) The allometric interpretation of the self-thinning rule. *Journal of Theoretical Biology* **89**: 475–500.

White, P., and Schwarz, A. (1998) Where do we go from here: the challenges of risk assessment for invasive plants. *Weed Technology* **12**: 744–751.

Whittaker, R. H. (1970) *Communities and Ecosystems*. London: Macmillan.

Whittaker, R. J. (2000) Scale, succession and complexity in island biogeography: are we asking the right questions? *Global Ecology and Biogeography* **9**: 75–85.

Wiegert, R. (1962) The selection of an optimum quadrat size for sampling the standing crop of grasses and forbs. *Ecology* **43**: 125–129.

Wiens, J. A. (1995) Landscape mosaics and ecological theory. In *Mosaic Landscapes and Ecological Processes*. Hansson, L., Fahrig, L., and Merriam, G. (eds). London: Chapman & Hall, pp. 1–26.

Wilcove, D.S., Rothstein, D., Dubow, J., Phillips, A., and Losos, E. (1998) Quantifying threats to imperiled species in the United States. *BioScience* **48**: 607–615.

Williams, G. (1975) *Sex and Evolution*. Princeton, NJ: Princeton University Press.

Williams, J. A., and West, C. J. (2000) Environmental weeds in Australia and New Zealand: issues and approaches to management. *Austral Ecology* **25**: 425–444.

Williamson, M. (2002) Costs and consequences of non-indigenous plants in the British Isles. URL www.ou.edu/cas/botany-micro/ben/ben281.html

Williamson, M., and Fitter, A. (1996a) The varying success of invaders. *Ecology* **77**: 1661–1666.

Williamson, M. H., and Fitter, A. (1996b) The characters of successful invaders. *Biological Conservation* **78**: 163–170.

Willis, A., and Blossey, B. (1999) Benign climates don't explain the increased plant size of non-indigenous plants: a cross-continental transplant experiment. *Biocontrol Science and Technology* **9**: 567–577.

Willis, A., Thomas, M., and Lawton, J. H. (1999) Is the increased vigour of invasive weeds explained by a trade-off between growth and herbivore resistance? *Oecologia* **120**: 632–640.

Willis, A., Memmott, J., and Forrester, R. (2000) Is there evidence for the post-invasion evolution of increased size among invasive plant species? *Ecology Letters* **3**: 275–283.

Wilson, E.O. (1985) The biological diversity crisis: a challenge to science. *Issues in Science and Technology* **2**: 20–29.

Wilson, E. O., and Peters, F. M. (eds) (1988) *Biodiversity*. Washington DC: National Academy Press.

Yarranton, G. A., and Morrison, R. G. (1974) Spatial dynamics of a primary succession: nucleation. *Journal of Ecology* **62**: 417–428.

Young, B. (2000) Invasive species and other priorities for English Nature. In *Exotic and Invasive Species: Should we be Concerned?* Bradley, P. (ed). Proceedings 11th Conference of Inst. of Ecology and Environmental Management: Institute of Ecology and Environmental Management, pp. 2–11.

Zavaleta, E., Hobbs, R., and Mooney, H. (2001) Viewing invasive species in a whole-ecosystem context. *Trends in Ecology and Evolution* **16**: 454–459.

Zavaleta, E. S. (2000a) Valuing ecosystem services lost to Tamarix invasion in the United States. In *Invasive Species in a Changing World*. Mooney, H.A., and Hobbs, R. (eds). Washington, DC: Island Press, pp. 261–300.

Zavaleta, E. S. (2000b) The economic value of controlling an invasive shrub. *Ambio* **29**: 462–467.

Zavaleta, E. S., and Royval, J. L. (2002) Climate change and the susceptibility of U.S. ecosystems to biological invasions: two cases of expected range expansion. In *Wildife Response to Climate Change*. Schnieder, S. H., and Root, T.L. (eds). Washington, DC: Island Press, pp. 277–341.

Zeevalking, H., and Fresco, L. (1977) Rabbit grazing and diversity in a dune area. *Vegetatio* **35**.

Zwölfer, H., and Harris, P. (1984) Biology and host specificity of *Rhinocyllus conicus* (Frol.) (Col. Curculionidae), a successful agent for biocontrol of the thistle, *Carduus nutans* L. *Zeitschrift für angewandte Entomologie* **97**: 36–62.

Zwölfer, H. (1973) Competition and coexistence in phytophagous insects attacking the heads of *Carduus nutans* L. In *Proceedings of the II International Symposium on Biological Control of Weeds*, pp. 74–80.

Zwölfer, H. (1985) Insects and thistle heads: resource utilization and guild structure. In *Proceedings of the VI International Symposium on Biological Control of Weeds*. Delfosse, E. (ed). Ottawa: Agriculture Canada, pp. 407–416.

Index

Page numbers in **bold** type refer to tables; those in *italic* type refer to figures and boxes.